Evolution Now

Evolution Now

DAVID·PENNY

Library of Congress Control Number:		2017901796
ISBN:	Hardcover	978-1-4990-9931-7
	Softcover	978-1-4990-9929-4
	eBook	978-1-4990-9930-0

Print information available on the last page.

Rev. date: 05/04/2017

To order additional copies of this book, contact:
Xlibris
0-800-443-678
www.Xlibris.co.nz
Orders@Xlibris.co.nz
753217

Contents

Preface

Good Science, never believe all our ideas.

This book is written from a Popperian perspective. Although Karl Popper (1902–1994) grew up in Vienna in Austria and had written some early works there, he had spent the Second World War years teaching philosophy in Christchurch, New Zealand, and he wrote his two-volume book *The Open Society and Its Enemies* (1945) whilst there. As a philosopher of science, he obviously (but perhaps unexpectedly?) had had a lasting influence on the scientists at that university. I did my undergraduate education there over a decade after Karl Popper had left for a position in the London School of Economics. For example, the older chemists might be giving a lecture on chemistry; and then they would change tone just a little and say that you were never to 'believe' your hypotheses but to test them, test them, and still test them again. Then they would revert to their original tone and go back to teaching chemistry. I guess that most of the undergraduates did not pick up that they were being given a Popperian interlude. But because I was also taking a philosophy course for science students that had one lecture a week on the philosophy of knowledge, I recognized it as a Popperian interlude in the chemistry class. So as a philosopher, he had had a major influence on the scientists there, much more so than most philosophers.

The early twentieth century had been a very important time for understanding science. For over 200 years, people had 'believed' that

Isaac Newton's laws to be true, absolutely true, and true for all time. And then along came Albert Einstein with his theories of relativity, and suddenly Isaac Newton's 'laws' were just very good approximations—at least at lower velocities. What was going on? Karl Popper was a philosopher who took science very seriously (e.g. Popper 1959) and his interpretation was that

> humans had no special abilities to only believe/accept fully correct hypotheses,
>
> nor had we considered (formulated) and tested all possible hypotheses,
>
> the best we could achieve was to keep testing hypotheses,
>
> it is easier prove hypotheses wrong, or inadequate, but
>
> to accept and use the most thoroughly tested hypothesis at any time.

So in a sense, it is an optimistic approach to science—we should never 'believe' our current ideas, but there is real progress with time. We have never yet thought of nor tested all possible hypotheses—we might be missing some important point. And so in this vein, we will start at the beginning of the seventeenth century—we are all still learning, and none of us know what the future might bring. But although there is some real progress in science, none of us know the future, but it is an optimistic future (see his autobiography, Popper 1976).

So yes, under the Popperian theme, we can make good progress, but we should never (ever) 'believe' our current models—just use the most thoroughly tested of them. Once we apply these principles to evolutionary theory, we can make some real progress in understanding it. I do think that the Popperian model does need to consider the subdivision of hypotheses; if a hypothesis consists of up to twenty subhypotheses (Chapter 3), it may be unclear which (if any) are not correct! It is important that we continue to test (and extend) our favorite ideas. Some of our ideas may be wrong; others will be extended in the future. No, we don't know which ones yet—which will be wrong, which will be updated and/or modified or extended. But this certainly does not stop us from trying to make some guesses (and in some later chapters, I will give some suggestions where I think we are wrong—or at least uncertain). So each of these will be illustrated—good progress, future questions, and some possible errors.

So we will start in the early 1600s and see that over the next 100 years, there was very good progress on several issues. In this introductory chapter, we will consider just two main themes that are covered by evolution;

> rejection of the older idea of continued spontaneous generation
> (the rise of the creationist species concept!), and
> the acceptance of science as having authority in selected areas.

Oops, perhaps we need also to consider the 'species' concept; it (as we show) is a pre-evolutionary idea. There is 'something for everyone' in this book; some ideas (towards the end) will be disputed by some people. However, science is an ongoing process and welcomes real challenges.

CHAPTER 1

Continuous Creation: European Indigenous Knowledge

From Continued Spontaneous Generation to the Fixity of Species Concept

The seventeenth century was a triumph for the early science of biology; careful observation and experiment started replacing what we will call European indigenous knowledge. Okay, there had been an earlier start since the times of the ancient Greeks, but we will start from the relatively neglected biology of the 1600s (the seventeenth century). Under the ideas from the earlier Middle Ages, there was a mixture of observations, miracles, and strange tales. From an earlier generation, the book of Ambroise Paré (1510–1590) from the late 1500s (sixteenth century) is a very good example (Paré 1585). He has interesting tales of imaginary creatures—demons that take human form, monsters that appear half human and half dog, and so on. An incubus was a demon in male form, a succubus in female form—both apparently had insatiable sexual appetites. It is interesting that Paré gave the source of many of his stories, for example, a three-headed lamb born in a village in 1577 and the name of the surgeon who had verified the information. Nevertheless, trying to authenticate the data is the beginnings of science.

So that certainly was a good start, giving the source of his information. Other stories referred back to the time of Saint Augustine (an early Christian scholar) or back to the Old Testament or to other sources. For example, the Old Testament gives an example of the 'inheritance of acquired characters'. Genesis 30: 37–39 reads as follows: '*Then Jacob took fresh sticks of poplar and almond and plane trees, and peeled white streaks in them, exposing the white of the sticks. He set the sticks that he had peeled in front of the flocks in the troughs, that is, the watering places, where the flocks came to drink. And since they bred when they came to drink, the flocks bred in front of the sticks and so the flocks brought forth striped, speckled, and spotted.*' Easy, isn't it? If it was true!

During the seventeenth century, we will see that early biologists went from an acceptance of continued spontaneous generation to the idea that species had a permanence in nature and through time. Today, we come back a little from this position and have 'species' evolving over time—but the idea of 'continuous spontaneous generation' of multicellular plants and animals had certainly gone by about 1700 from science (the beginning of the eighteenth century). It is interesting to read comments about work in the early 1600s, in this case about the animal *Virgularia*.

> *Captain Lancaster, in his voyage in 1601, narrates that on the sea-sands of the Island of Sombrero, in the East Indies, he 'found a small twig growing up like a young tree, and on offering to pluck it up it shrinks down to the ground, and sinks, unless held very hard. On being plucked up, a great* worm *is found to be its root, and as the tree groweth in greatness, so does the worm diminish; and as soon as the worm is entirely turned into a tree it rooteth in the earth, and so becomes great. This transformation is one of the strangest wonders that I saw in all my travels: for if this tree is plucked up, while young, and the leaves and bark stripped off, it becomes a hard* stone *when dry, much like white coral: thus is this* worm twice transformed into different natures. *Of these we gathered and brought home many'.*

In this extract from Charles Darwin's *The Voyage of the Beagle* (1845), he comments on ideas from 1601. At that time, it was assumed (as part of European indigenous knowledge) that matter could readily transform between 'animal, vegetable and mineral'—that is, between three very 'different natures'. In this case, the organism appeared to change between all three—animal, vegetable, and mineral! Figure 1.1 has a picture of *Virgularia*.

This extract is a good example of thought at the beginning of the seventeenth century—matter could change between animal, vegetable, and non-living mineral. Read the extract again; Captain Carpenter *observed* the changes between 'animal, vegetable and mineral'. This European indigenous knowledge of the early 1600s is our starting point; but later more careful scientific study, including testing better hypotheses and more careful observation and testing, led to our improved understanding. Thus we outline the understanding of the 'forms' of plants and animals at the beginning of the seventeenth century (1600s) and then examine the remarkable changes that occurred in biology during that century, especially the development of the idea of 'species' having a permanence through time. That is a good start—though certainly not the final word!

Fig. 1.1. *Virgularia*, a sea pen.
This belongs to the phylum Cnidaria (jellyfish, sea anemones, corals, and hydroids). Class: Anthozoa, Subclass: Octocorallia (sea pens, sea fans, and soft corals). It forms elongated, very slender colonies with narrow 'leaves' bearing polyps. A European species is up to 600 mm long with central stem only a few millimeters thick. All sea pens possess an internal calcareous skeletal rod (axis); in this species, it is round in section and often protrudes from the top of the colony. It lives embedded in mud, into which it can withdraw.

Often there is some grain of truth behind the stories. For example, legends about unicorns are assumed to derive from finding subfossil remains in Europe of a lightly built rhinoceros that had a single horn. Given that horses are about the closest relative of rhinoceroses, their skulls could be interpreted as a 'horse-like creature with a single horn'—a unicorn. Thus there may be some knowledge around 'beliefs'. So in this chapter, we will focus on the major advances in going from ancient beliefs that spontaneous generation continued to occur to the modern view that there is no continued spontaneous generation (nor is there continued inter-conversion of living and non-living forms of multicellular plants and animals). It was not really explained (at that time) why any species that occurred in the past no longer existed (they could now live somewhere else on earth!). Later we will look at the origin of life—are there any really good (testable) theories? Yes, so let's test them!

Although biology made many major advances during the seventeenth century, we focus here on the change from the European traditional (indigenous) idea of *continued* spontaneous generation to that of species having a permanence throughout time. The original view in Europe was that 'kinds' or 'forms' of multicellular plants and animals continued to arise by spontaneous generation. In the words of John Ray ((1627–1705) who contradicted this, see later), the work of Creation was *not* considered finished—it continued into present times. The Creator was all powerful and was not bound by what s/he had done in the past. (Okay, I never know what gender God is - or was). The early gods were apparently female (earth mothers). But then in Hinduism, they are both male and female; and in the Judeo/Christian/Muslim tradition, God becomes male. So I just use the phrase 's/he' in order to be neutral! It is pronounced sha/he or shu/he.) It was also accepted that there was continued inter-conversion between living forms and between living and non-living matter. After all, if a caterpillar could turn into a stone (chrysalis) and then into a butterfly, why couldn't a barnacle turn into a goose or a plant into an insect? The Old Testament was interpreted to provide support for continued spontaneous generation:

And God said, Let the earth bring forth grass, the herb yielding seed, and the fruit tree yielding fruit after his kind whose seed is in itself, upon the earth: and it was so,
And God said, Let the earth bring forth the living creature after his kind, cattle, and creeping thing, and beast of the earth after his kind, and it was so. (Genesis 1: 11, 24)

There was nothing here to indicate that this power of the Creator to generate living forms had suddenly ceased. The passages were interpreted as accepting continued spontaneous generation; spontaneous generation did not occur just once—it continued to happen again and again and again. This was just part of European indigenous knowledge. Indeed, it seemed to be a heresy in Europe during the early Middle Ages not to accept that spontaneous generation still continued (see later). That would be to deny the abilities of an all-powerful Creator to design a system that continued to generate life. It is important to know how widespread these indigenous beliefs were; they existed in much of the rest of the world in addition to Europe and the Middle East (Farley 1977; Oparin 1968). Most of the world's cultures appear to have ancient stories about the transformation between different 'forms'. A story from the Admiralty Islands of Papua New Guinea tells of a giant turtle laying eggs—eggs that turned into humans. So that was how humans arose—case solved!

Also note that there is no reference to 'species' in these quotations, just to 'kinds' or 'forms' of plants or animal. The concept of a biological species did not develop until much later in the seventeenth century (see later, 1691 in particular), and the above translation into English authorized by King James I is from early in the century (1611). As mentioned above in the quote from Captain Lancaster, there appeared to be ample observational evidence for an object transforming between 'animal, vegetable and mineral'. This introductory extract illustrates how in Lancaster's time (1601) it was just accepted that one form of matter could be transformed into another—in this case, between vegetable (a plant), animal (a worm), and mineral (a coral-like 'stone', apparently non-living). But now we know better.

At the time of *The Voyage of the Beagle* (230 years after Lancaster), there was a general acceptance that a species had a continuity over many generations and that they did not transform into other forms (or kinds or species) over the lifetime of individuals. But at the beginning of the seventeenth century, there was no firm division into living and non-living nor between animal and vegetable. Indeed, if caterpillars could be transformed into a 'rock-like' material (a chrysalis) and these turn into butterflies, then couldn't human tissue make the apparently simpler transformation into intestinal worms or into liver flukes? It required a large amount of careful observation and experiment to determine the continuity of each form of life over many generations—careful observation and experiment that we now take for granted.

The belief in spontaneous generation is much earlier than the early seventeenth century, and it can be found in the writings of the ancient Greeks such as Aristotle (about 330 BC) (see the first chapter in particular of Oparin 1968). Aristotle was the most accomplished zoologist of his time and recorded (and checked wherever possible) everything that was known about animal life. For 2,000 years, his knowledge appeared to dominate European views on many topics, including zoology. Much of what Aristotle recorded probably came for earlier and widespread beliefs, and certainly the passages from Genesis are older again. So we need to accept that we, as humans, keep learning.

An example of a mid-seventeenth-century belief is a recipe for the spontaneous generation of mice. The distinguished Brussels physician (and chemist and physiologist) Jan Baptist van Helmont (1580–1644) helped make important advances in the understanding of plant nutrition. For example, he showed that the increase in weight (growth) of young willow trees did *not* come from the soil (because the soil had lost very little weight, just a few ounces). Thus the increase of seventy-four kilograms in matter of the tree must come from water (H_2O); nowadays we would also say carbon dioxide (CO_2, which he also discovered). That was a major advance over the concept of the soil turning into plant (Reeves et al. 1999: 77).

Nevertheless, Van Helmont's experiments on the origin of mice showed an acceptance of spontaneous generation: *take grains of wheat and a sweaty shirt, place them under a box in a field, and leave for three weeks*' (Isley 1994). The vapors of the shirt, together with the vapors of wheat, would generate live mice, which Van Helmont reported were very similar to ordinary mice! Just think about it—you could try the experiment for yourself, and it would almost certainly work (provided that your shirt was sweaty enough!). So Van Helmont's experiment was based on 'observation'—just not careful enough by our standards. Indigenous knowledge needs to be taken seriously, but not 'believed'— just tested. Thus Van Helmont's contribution to plant biology is taken for granted. It is especially interesting to see it as a step towards our modern understanding (but what will we learn in the future?).

There are many other examples. For example, eggs of the barnacle goose could not be found, and a common interpretation was that some invertebrates on the rocky shore turned into geese in the spring. The name 'barnacle' is now associated with these marine invertebrates—the supposed progenitors of the barnacle geese. It was assumed in some parts of Europe that melons turned into lambs; therefore the eating of lamb on Fridays was acceptable (in regions where eating meat on Friday was unacceptable on religious grounds). Similarly, in Paris, a bird (a macreuse) was allowed to be eaten as fish during Lent, and the legend of the geese from barnacles was used to justify that the macreuse really came from fish. An early English naturalist (John Ray, see later, and also Raven 1986) asked visitors to Paris to send him skins from the bird, and he identified it as a form of sea duck, the scoter. Unfortunately (for eating meat during Lent), its reproduction was known, and another example of the 'transformation' of one living form to another was invalidated.

Near the middle of the seventeenth century (1668), Italian researcher Francesco Redi ((1626–1697) Hawgood 2003) carried out experiments to demonstrate that maggots did not arise spontaneously from meat. Redi knew that butchers covered their meat with muslin cloth. The butchers knew that under these circumstances, maggots did not form. By careful observation and experiment, Redi was able to show that maggots developed from eggs laid by flies; thus the maggots did not develop spontaneously from meat—there was a life cycle of flies, eggs, maggots, flies, eggs, maggots, flies, eggs, and so on. Redi's experiments were repeated throughout Europe and were important in challenging the belief in continued spontaneous generation. However, there was still an easy fallback position: 'maybe dead-meat cannot be transmuted into new living forms, but living tissue can change into new forms' could paraphrase one counter-argument. Redi's results were challenged by several others in Europe. But the basic result stood: there were no cases of meat turning into maggots and then into flies; the flies laid eggs—dammit.

Two similar examples follow. Observation had shown that insects could emerge from certain swellings (galls) found on many plants (such as oaks). This was interpreted as living tissue of plants turning into insects. But more careful observation and experiment in the Netherlands by Jan Swammerdam (1637–1680), an early microscopist, found that the galls themselves were formed after the insects laid eggs on the plant material, and these eggs developed into insects. Again, there was continuity of 'species'—insects giving rise to insects and plants giving rise to plants, but not a transformation of a plant into an insect (Cobb 2000).

Similarly, from observation (and personal experience), it was known that many people and many animals harbored a variety of internal parasites, from tapeworms to liver flukes. This could be interpreted as one form of life (human or other animal tissues) transforming into another form of living matter (the internal parasite). It took many decades of careful work by many researchers to demonstrate the life cycles of all these parasites, that there was a continuity of like forms over many generations. It is important to separate multicellular plants and animals from unicellular forms only seen under the new microscopes (Farley 1977).

As mentioned earlier, it is not just Europeans who thought along these lines of continued transformation between forms of life; there are many stories from other parts of the earth. In New Zealand, lizards (and the tuatara) traced their ancestry in Polynesian mythology back to Tangaroa, god of the oceans. Seeking to escape the wrath of Tawhirir-matea (father of wind and storms) for the separation of Rangi (earth) and Papa (sky), Tu-te-wehiwehi and his descendants fled from the sea to the land and adopted lizard-like forms (Andersen 1969). In parts of Melanesia, there are stories of giant turtles laying eggs that developed into humans (see earlier). Early Hindu writings include similar stories of transformation between living 'forms'. Thus 'beliefs' about continued spontaneous generation appear to be much more general that just European.

It is often assumed that if a people have a 'name' for a plant or animal, then that is equivalent to having a concept of species. The examples above show that this is not correct; having names for plants and animals is certainly not sufficient for this conclusion. More than one name could be associated with a single biological species; a classic example is names being given to each age class of some freshwater fish. A fish could be a smelt, then sprod, mort, forktail, half fish, and finally a salmon. In the seventeenth century, biologists such as John Ray dissected them and found that only the largest had sperm and ova; thus only the largest form could reproduce (Raven 1986). So Ray concluded that the names referred to different age classes, all belonging to the same 'species'. Thus it is time to consider the idea of species that developed in the latter half of the 1600s.

Given this background (and with animals and plants showing developmental stages), the scientific hypothesis of 'like from like', or 'the fixity of species', was a major and important scientific advance during the seventeenth century. It replaced these earlier ideas that 'forms' or 'kinds' of life (and living and non-living matter) were continuously being transformed. If 'ice' could form spontaneously at low temperatures, why could 'life' not form spontaneously? The concept of species as we know it today developed in the latter half of the seventeenth century, and of course, it was only after the development of the species concept that the question of the 'origin' of species became a valid problem. In 1651, when William Harvey (1578–1657), who earlier had discovered the circulation of the blood, published his book on *Exercitationes de generatione animalium* (*Experiments Concerning Animal Generation*) (Harvey 1651), it was becoming clear that in mammals, generation followed generation followed generation. Although Harvey found that an individual was always born of like parents, he still could not show full continuity from generation to generation; to show complete continuity between parent and offspring of a mammal would require at least a microscope to be able to see egg and sperm. The phrase 'all life from eggs' (or *omne vivum ex ovo* in Latin, the main scientific language of the time) is often used to describe Harvey's work. Harvey still interpreted 'egg' very broadly, and it could still allow spontaneous generation of worms, insects, etc. As more studies were done, the phrase 'all life from eggs' became a common way of denying the transformation of living tissue into new forms. It was extended to also express the conclusion of the fixity of species.

An influential figure in the development of the species concept was the naturalist/cleric John Ray (Raven 1986). He is perhaps mostly remembered today for his work on classifying plants, but he worked extensively also on birds, fish, and insects and was one of the early writers on natural theology (see later). Before classifying plants, he wanted to know as much as possible about them; wherever possible, he collected plants (or seeds) from different environments and grew them at his home under the same conditions, studying their seeds, their germination, and their progeny. Perhaps with plants it was much easier to show the continuity of matter from one generation to another. The fruit and seeds were quite obvious, and the germination or growth of the young plants was easy to observe. He found that some 'named varieties' could grow from seeds from the same plant. Consequently, he considered them members of the same species—so that is a good start.

His concept of a species can be summarized as a group of organisms who were '*born to members of the group, mate with members of the same group, and gave birth to members of the same group*'. For example, '*I reckon all Dogs to be of one Species they mingle together in generation, and the breed of such mixtures being prolifick*' The latter is particularly interesting because the size and shape variation on the species 'dog' is greater than between many genera in nature. In these early days, purely morphological differences were not sufficient to describe forms as distinct species.

John Ray's concept of a species from 1691 (see later) is virtually the biological species concept as we know it today. If a crossing results in fully fertile offspring, then the parents are generally considered to be of the same species. Using this concept, Ray insisted that some 'species' that had been described by earlier workers were just normal variants in the population—because different variants could appear as offspring from a single plant. Similarly, he transplanted plants from different regions to his own garden; this meant he could detect some environmental effects, thus separating heritable (genetic) from environmental changes. Some people today (300 years later) don't even do this elementary precaution (transplanting to the same site), let alone grow the plants and make crosses, before pronouncing a plant to be a new species. In some areas, we appear to have regressed over the last 300 years. Surely not!

Remember that the quote from Genesis 1 uses 'kind' rather than 'species'. When the Bible was originally written and much later when it was translated into English in the early 1600s, there was no biological species concept as we know it. Some people have in the last 150 years (after spontaneous generation was widely rejected) interpreted 'kind' as equivalent to 'family' or 'order'; this would allow them to accept a fairly large amount of evolution—there could be creation of a limited number of forms or types and then evolution produced, say, to families, genera, and species.

Ray's species concept is from his book on natural theology, *The Wisdom of God Manifested in the Works of Creation* (Ray 1691), and this is from where we take his species concept. It is apparently the first popular book (that is, written in English, not in Latin—the main scientific language of the time) that described this new view of species. Species now had a permanence through time, and there was no continued spontaneous generation of complex life forms. But Ray's book was initially criticized on religious grounds for denying the importance of continued spontaneous generation (as interpreted in the Genesis quote given earlier). In his second edition (in 1692), he included an apology if he had offended readers by his 'confident denial' of continued spontaneous generation, and then he explained why he thought species had a long-term existence. Note that this was (in our concept) a religious book that defined 'species'.

I wanted to check that second edition's 'confident denial' of continued spontaneous generation, but where was I to find a book in New Zealand published in 1692? The universities didn't start until 1869, nearly 180 years later (and it was before early books were available on the Web). I was visiting Cambridge University the next week, and so I asked friends to see if the book was available there. (John Ray had been thrown out of Cambridge University when the monarchy was re-established in 1660; he was willing to accept back the monarchy, but not the persecution of dissenters. Only later when William and Mary came onto the throne (in 1689) did he consider it safe to write his book on natural theology.) Yes, Cambridge did have copies of the second edition, but a friend advised against trying to see a copy in a college library (it would take months to get through the college bureaucracies, he said), but the Botany Department library has a copy! Yes, it certainly had the 'confident denial' of continued spontaneous generation (and in a book on natural theology!).

At this same time when spontaneous generation was being questioned and species were being considered stable, early chemists (possibly alchemists) were denying that elements could be chemically transmuted (denying that mercury could be changed into gold, for example). Thus there was a simultaneous movement in scientific thought:

> denying transformation of non-living forms of matter (chemical elements),
> denying spontaneous generation (non-living to living), and
> denying transmutation between living forms (species therefore having some permanence through time).

Yes, we now know that chemical elements can form from simpler chemical elements when they crash together at high velocities—the original universe was mostly hydrogen, and the other elements formed later. However, that does not affect the general idea.

Thus the development of the concept of species was an important step in the rise of modern science. It allowed three important scientific advances:

> continued spontaneous generation (of multicellular forms of life did not occur, but there were still microbes),
> members of a species were an interbreeding group, and
> species had a permanence (or stability) through time.

We accept the first two today and agree that most species have some stability in at least the shorter term of 100 or 1,000 years or so. But in the seventeenth century, it was a major scientific advance though to suggest that most species would have an existence for at least 6,000 years. Today we go much further and see that a few species (the tuatara (*Sphenodon*) is one example, the horseshoe crab (*Limulus*) and the maidenhair tree (*Ginkgo*) are others) have existed for tens of millions of years. It was the discovery during the late eighteenth and early nineteenth centuries of a much longer timescale (an older earth and older solar system) that made the important step towards our modern understanding of the age of the universe (this comes later in Chapter 2).

Nevertheless, it is a little amusing to think that Ray had to apologize to some of his (religious) readers because he so strongly accepted that species had a permanence through time. Similarly, it is a bit humorous to think that some of our current 'born-again' creationists would have been burnt at the stake for being heretics in earlier (pre-1650s). But none of us know what the ideas and hypotheses in the future will be; we just have to follow the evidence as it stands. (I guess I can't help it. I am a little Popperian; we should 'use' our best tested ideas, not 'believe' them.) However, there is such a huge amount of evidence now that taxa evolve through time that we would certainly be surprised if that is overthrown. Evolution now seems inevitable, dammit; we can't stop evolution—those damned little RNA flu viruses just keep evolving (we make a different immunization every six months—for the northern and the southern winters)! But we certainly are getting closer to good testable models.

One additional comment on the species concept as it arose in the late seventeenth century is interesting. Perhaps we can see now that there were four sources of ideas that made up that early species concept. These might be:

1. the increased acceptance of the continuity of multicellular species between generations (all life from eggs),
2. an interbreeding group (the biological concept of species),
3. the idea of God as a creative force having ended (initially having created each species), and finally,
4. the idea from ancient Greece (e.g. the philosopher Plato) of unchangeable 'essences'.

It is the last two that cause a clash with modern science, but let's focus on the last. Plato (in ancient Athens) had argued that all 'forms' had an ideal 'essence' that was unchangeable. When this (pre-Christian) concept was applied to biological species, the idea that species were also unchangeable seemed a natural assumption. But in the longer term, this was the real clash with modern biology, and it is unfortunate that this Greek concept had gotten mixed up with the seventeenth-century concept of species. It might not be the Christian part that caused the main difficulty with this early (non-evolutionary) species concept, but it could be the Greek part. Anyway, it is an interesting thought—and it fits with the Popperian idea that we are learning as we do more measurements and observations.

The words 'genera' (singular form of 'genus') and 'species' originally came from Aristotelian logic (also from ancient Athens), and in modern terminology, they meant more or less a 'set' and a 'member of the set'. In this original meaning, something could be both a species and a genus, depending on the context. For example, John could be a species of the genus male, males a species of the genus human, humans a species of the genus animal, animals a species of the genus living creatures, and so on. This more original usage is found today in the words 'generic', 'general', and 'specific'. Although we now use 'species' almost exclusively to refer to biological species, there are dozens of other usages of the word 'species' listed in the full *Oxford Dictionary* (a 'species' of coin, for example), and most usages predate the biological usage that developed in the late seventeenth century.

As an aside, Carl Linnaeus, who formalized biological classification in the next (eighteenth) century, may have used a more Aristotelian definition of species, replacing the biological species concept (part 2 in the list above) with the concept of a group sharing characters. This made it easier to define species in the museum or in the herbarium without having to do breeding experiments. Nevertheless, Linnaeus was strongly against splitting species into smaller and smaller units just because some minute difference was apparent. In addition, and for plants at least, he considered that plant genera (not just species) may have been created (mentioned in Lockhart et al. 2014), and then natural processes led to new species, which we identify.

An interesting and important point about the past arises as a consequence of European indigenous knowledge from before the concept of species existed. Would you 'blame' early Europeans for hunting the Irish elk to extinction if they believed that the elk's 'kind' or 'form' could in some unknown location continually be regenerated? Of course, we don't know all the details of their particular form of indigenous knowledge, but it would have to be pretty effective in order to survive in Europe during the last Ice Age. Then even 100 years after the development of the species concept (say, up to about 1800), most scientists accepted that species could not go extinct because they were ideal forms made from an ever-wise creator. It was impossible, given their knowledge of the time, for a form, kind, or species to go extinct; and there were other parts of the world still to explore!

What about these other parts of the world? Much of the megafauna in the Americas, which had survived many major climate changes during the series of Ice Ages, disappeared within a short time of human arrival. In the last few years, it is apparent that K-selected species (that is, long-lived, slow-reproducing species) are particularly susceptible to overhunting and extinction. But we have only learned that recently. Should we 'blame' these early Indians for overhunting these species? Some early North Americans (Murray 2003) believed that bison bred in huge numbers elsewhere, disappearing each winter to breed in large numbers in pastures beneath the sea. Northern Europeans assumed that some birds (now known to be migratory) overwintered in the mud at the bottom of lakes and ponds, a belief that explained their disappearance during winter. Some Italian people believed that a god (pantheistic or monotheistic) sent them migratory birds every autumn especially for them to hunt and kill during late autumn in order to stock up for the winter period. They should show their gratitude by catching as many as possible.

Similarly to Europe and the Americas, in New Zealand, moa were a K-selected species and disappeared quickly after human settlement (Allentoft et al. 2014). We do need to say here that there may be reasons other than direct human hunting that might have caused extinctions (Hurles et al. 2003). In this publication, we give two other types of factors that might be important in specific cases: (1) habitat modification (including the use of fire, especially by early humans) and (2) the introduction (deliberate or accidental) of new plants, animals, and diseases—these can be considered to be invasive species like humans.

On other occasions, some local belief may have contributed to survival of a species. The 'tragedy of the commons' refers to the problem that if nobody (or no group) owns a resource, then people tend to over-exploit it, more or less on the grounds if we don't get that elk, then those people in the next village will, and then we lose out anyway. In England, swans and some deer may have survived because they were 'owned' by the monarch—consequently, swans survived (it was far too risky catching a swan if you got hanged when caught). A similar thing happened in some South Pacific islands; megapode birds (similar to the Australian brush turkey) survived when they belonged to the great chiefs. In contrast, in many Pacific Islands, the birds became extinct after human arrival.

Thus as mentioned earlier, in the early Middle Ages, modern-day 'creationists' would be considered heretics (and burnt at the stake) for denying the continued spontaneous generation of living matter. In this earlier time, people understood the Bible to imply continued spontaneous generation. It would seem an affront to the Creator not to be able to make something that could change. Think about it. An all-powerful Creator cannot make species that evolve. Why? Because Plato says that essences are unchangeable. But hold on. Plato was a pre-Christian pagan—why should an all-powerful Creator be limited by an ancient Greek philosopher? On this argument, modern creationism must be a heresy! But please don't tell anyone—and ideas do evolve with time.

However, we now have a concept of species as opposed to continuous spontaneous generation of multicellular plants and animals. It took another 150 years before spontaneous generation was eliminated by Louis Pasteur (1822–1895) for microbes as well, but the concept of species was an advance at the time—and we come back to it later. It did go too far by considering each species was created. But now we should consider another major change that occurred during the seventeenth century.

Natural Theology and the Realm of the Scientists and
the Realm of Theologians

Around the same time (late seventeenth century), there was a strong movement (including Isaac Newton and Robert Boyle (of Boyle's law of gases)) that advocated that science should only study how things worked and not study the origin of things (in Newton's case, the origin of the solar system). Indeed, Greene (1959) refers to this view (science as not studying the 'origin of things') as 'Newtonian creationism'. Good scientists should not spend their time in idle speculation on how things arose was apparently the reasoning. Real scientists should study how things worked in the present when actual evidence could be obtained! The origin of solar systems, or the origin of species, was the realm of the Creator; scientists should study the principles on how these systems worked. But try and tell that to a modern physicist who studies the origin of solar systems!

This division into how the world worked and how it 'arose' was an apparent early division into the 'realm of science' and the 'realm of theology'. It was probably initiated earlier in the seventeenth century by the scientist/philosopher René Descartes (1596–1650) when he suggested both that there was 'physical matter' (that was the realm of scientists) and 'mind matter' (which was the realm of the all-powerful theologians, who would happily burn you at the stake if your experiment got the wrong answer—well, if you were crazy enough to tell anybody about your results). Similarly, Descartes advocated that scientists were to study existing processes (how God's law worked) and not study the questions of origins—that was God's business!

For the question of 'mind matter', Descartes (1637) says, '*There is none that is more powerful in leading feeble minds astray from the straight path of virtue than the supposition that the soul of brutes is the same nature with our own*'. Slightly earlier, he says (of, for example, the Great Apes), '*remain two most certain tests . . . Of these the first is that they could never use words or other signs arranged in such a manner as is competent to us in order to declare our thoughts to others* (p. 56) . . . *did not act from knowledge* . . . (p. 44) *for while reason is a universal instrument that is alike on every occasion, these organs, on the contrary, need a particular arrangement for each particular action'* (p. 45).

Yes, it was a brilliant compromise. On one hand, there was no evidence for this artificial distinction into two forms of matter. But it was still a good step for the time that the all-powerful theologians should no longer persecute the emerging group of scientists just for studying nature. There was an area where the early scientists were the official authorities. It had not always been that way, for example:

> Earlier (in the mid-1500s) Copernicus arranged to publish his views of the sun-centered solar system only after his death.
> Bruno had been burnt at the stake for (among other things) advocating the earth was not the center of the solar system.
> Galileo was threatened with torture (he thought the earth did go around the sun).

Raven (1947) describes how several early botanists (whom we think of as nice harmless people) were burnt at the stake, so these were not isolated cases. Pinker (2011) describes the 50–100 times reduction in violent deaths in Europe between about 1300 and 2000. So there is both social and scientific progress, and real progress has led to a major reduction in deaths by violence (maybe we should tell ISIS supporters).

Let's get back to René Descartes. In his compromise, science had its legitimate place—as long as it kept to things observable and did not consider either origins or the human mind. This compromise lasted for perhaps 150 years, but from the early nineteenth century, many scientists were interested in origins, whether it be the origin of the solar system or of chemical elements or of species. They all became questions in the realm of science. Thus although the Descartes compromise was an important step along the way of the rise of scientific study, it was just that—a step. We can summarize his ideas in Figure 1.2 and then consider the people who later broke down some of the boundaries. It took until the eighteenth and early nineteenth centuries before barriers to the scientific study of origins broke down for the physical and then for the biological worlds. For example, Pierre-Simon Laplace (1749–1827) showed mathematically that the solar system could arise from a massive rotating cloud of dust that would (just by Newton's laws) slowly aggregate into a central sun with planets rotating around it in an ellipse, all moving in the same plane and in the same direction. Perhaps it took even longer, until the latter half of the twentieth century, for some people to eliminate some barriers to the study of mind in humans and/or animals even though evolutionary biologists had studied it earlier (see later). Nevertheless, by the end of the seventeenth century, European scientists had their area of authority and did not expect to be burnt at the stake by overzealous theologians—as long as they kept to their restricted area. Not all regions of the world advanced so quickly.

	physical matter	mind matter
how it works (mechanisms)	Yes √ Newton	No Darwin, Yes
How it arose (its origin)	No Laplace & Darwin Yes	No Darwin, Yes

Fig. 1.2. The realm of science around the end of the seventeenth century. Science could study how physical matter worked (including biology), but not its origins and not the human mind. Later, Laplace, for example, showed that explanations for the origin of the solar system were possible; and later again, Darwin demonstrated both the origin of species and the human mind were within the realm of science.

It is important to consider the importance of this 'natural theology' compromise. We have already had an introduction to this with John Ray's book *The Wisdom of God Manifested in the Works of Creation*. Over 100 years later, William Paley (1743–1805) was still writing an influential book that followed a very similar approach. The following extract is a famous passage from the beginning of Paley's *Natural Theology or Evidences of the Existence and Attributes of the Deity* (1802). When Charles Darwin was an undergraduate at Cambridge University, he had to read Paley's *The Principles of Moral and Political Philosophy* because it was required reading for all Oxford and Cambridge undergraduates! The book on natural theology begins with one of the most elegant passages in written English:

> *In crossing a heath, suppose I pitched my foot against a stone, and were asked how the stone came to be there; I might possibly answer, that, for any thing I knew to the contrary, it had lain there for ever; nor would it perhaps be very easy to show the absurdity of this answer. But suppose I had found a watch upon the ground, and it should be inquired how the watch happened to be in that place; I should hardly think of the answer which I had before given, that, for any thing I knew, the watch might have always been there. Yet why should not this answer serve for the watch as well as for the stone? Why is it not as admissible in the second case, as in the first? For this reason, and for no other, viz. that, when we come to inspect the watch, we perceive (what we could not discover in the stone) that its several parts are framed and put together for a purpose, e.g. that they are so formed and adjusted as to produce motion, and that motion is so regulated as to point out the hour of the day . . . This mechanism being observed (it requires indeed an examination of the instrument, and perhaps some previous knowledge of the subject, to perceive and understand it; but being once, as we have said, observed and understood) the inference, we think, is inevitable, that the watch must have had a maker; that there must have existed, at some time, and at some place or other, an artificer or artificers who formed it for the purpose which we find it actually to answer; who comprehended its construction, and designed its use.*

Read it again; it is a delightful example of English prose. Charles Darwin said that he was very impressed with Paley's argument as an undergraduate but later described his *On the Origin of Species* as 'one long argument' against the reasoning of Paley. Darwin did propose an alternative that did not require an 'artificer'—but that comes later.

As mentioned, natural theology has been a strong theme (perhaps more so in English science?) from the time of René Descartes, John Ray, and others by the end of the seventeenth century. The premise of natural theology was that evidence about the Creator could be learned from studying nature—hence the term 'natural' as contrasted with theology as supposedly 'revealed' in the Bible or supposedly by 'direct communication'. There were plenty of scientists who initially accepted some type of 'theological evolution', where a Creator had some say in, for example, directing evolution or selecting which mutations might occur. These were important steps on the way to fully scientific (testable) models.

Thus the right of science to study the physical world (including biology) was well established before the end of the seventeenth century. The traditional view was that the physical world was imperfect and temporary and that it was immoral (or at least frivolous) to study it— people should study and consider only the perfect world that is to come in the future. Much of our early modern science of the seventeenth to nineteenth centuries appears to have grown out from a natural theology perspective. The study of 'nature'—astronomy, physics, chemistry, and biology (including humans)—was thought to help understand the universal laws the Creator designed. The physical world was worthy of study in its own right and operated in a predictable way by laws that were discoverable. In this sense, natural theology was an important step towards our current understanding of the universe—but only a step along the way.

Summary of Chapter 1

Perhaps we should now reflect on what was learned during the seventeenth century and why we consider it quite an advance for biology. But first, there is the small Popperian commentary that we should never 'believe' our models—just test them and use the best tested one at any one point in time. Humans have no abilities to only accept fully correct models, but under no circumstances are we to 'believe' our hypotheses and not test them. It is an optimistic model of science that allows fluidity and progress—though more on that later.

At the beginning of the seventeenth century, we start with the idea of continued spontaneous generation of all forms of life. We end up with a creationist species concept! But this is a good start, though it is still limiting us in many ways—and the species concept is still a pre-evolutionary idea. Perhaps some conservationists still think as if species really exist. And there are apparently many unreasonable current limitations for the genesis of new species, including hybridization. 'Species' really are dynamic entities, and we come back to this question.

And it is still amusing to think that a modern-day creationist could have been burnt at the stake earlier for denying the power of a Creator to continue creating life forms. Okay, maybe we shouldn't joke about such things; humans are always learning. And there are some things that even today where I suspect that a few of our current-day 'beliefs' are incorrect (or certainly incomplete)—more later.

But then there was the recognition of areas where scientists are the authorities. This was an advance, and although we would now call it 'limiting', it was a real advance at the time. The limitations had essentially broken down by the end of the eighteenth/early nineteenth centuries for the physical sciences and probably by the mid-nineteenth century for the life sciences—though perhaps there were a few holdouts until the late twentieth century over 'mind matter'. But still there was real progress during the seventeenth century.

CHAPTER 2

The Eighteenth and Early Nineteenth Centuries

So now we have both the concept of species and science having authority in limited areas. This is good progress, but now we move to some new areas and in particular consider,

> the necessity for a long geological timescale,
> the importance that Charles Darwin started his professional career as a geologist, and
> some early ideas of the relationship between species.

So these are the three areas that we focus on in this chapter. It was from geology that the idea of the age of the earth being many millions of years old was developed, and this was essential for a mechanistic approach that eventually led to evolution. Second, the significance of Charles Darwin starting his professional career as a professional geologist is not welcomed by some biologists, but it does seem important. And finally we consider some of the ideas that were suggested for the relationship between species—certainly not all biologists accepted that all species had been 'created' in their present form.

The Long Timescale—Millions and Millions of Years

By the beginning of the eighteenth century, the idea of continued spontaneous generation was largely abandoned for the larger multicellular plants and animals even though it was another 150 years before the experiments of Pasteur (in the mid-nineteenth century) eliminated it for microbes as well. Once the hypothesis of species had been widely accepted (existing for at least 6,000 years, from 4004 BC on the timescale of Bishop Ussher in 1658), it led to the two additional questions: 'Do species change with time?' and 'How are species related?' These questions could only be considered once it was established that there were interbreeding units (species) that could be maintained over thousands of years, and the question of the lifetime of the earth became critically important. So the first hypothesis for the origin of species was the separate creation of each and every species ('*God must have an inordinate fondness of beetles*' was the comment of one eminent twentieth-century biologist, J. B. S. Haldane). However, even at these early stages, there were some evolutionary ideas that there had been change with time, either 'ascending' from a primeval form or 'degenerating' from an original perfect creation. However, the first topic really comes from geology.

Before continuing with the researchers and the questions they were studying, we should consider a major conceptual difficulty. At the beginning of the nineteenth century, biology was stuck. As long as a scientific explanation had to be compressed into quite a short timescale, there were no good and testable scientific theories to explain the origin of species. Then main scientific argument against evolution—demonstrating that species had been stable for at least 4,000 years—would then no longer be valid. French researchers, in particular following Napoleon's conquests, had shown that, for example, the mummified ibises found in ancient Egypt had scarcely changed from modern ibises. This was probably still true even if the timescale was extended to several tens of thousands of years, such as by the eighteenth-century French scientist Buffon (Georges-Louis Leclerc, Comte de Buffon (1707–1788)).

The conceptual change about time came not from biology, but from geology. The immediate source was probably the Scottish geologist Charles Lyell (1797–1875), though he lived in London most of his adult life. His three-volume *Principles of Geology: Being an Enquiry How Far the Former Changes of the Earth's Surface Are Referable to Causes Now in Operation* was published from 1830 to 1833 and convinced many scientists that a long timescale in the order of many millions or hundreds of millions of years was available—but where did Lyell get his ideas from?

Lyell's theme—that former changes to the earth's surface had to be explained by the same mechanisms that act in the present—came from an earlier Scottish geologist, James Hutton (1726–1797). During the last two decades of the eighteenth century, Hutton had developed his approach of studying geological events that occurred before a written historical record existed. His *Theory of the Earth: With Proofs and Illustrations*, published in 1795, expounded the necessity of keeping to known causes or mechanisms and not invented fantastic and huge forces that had only operated in the remote past, including great floods that supposedly covered the earth by some unknown process. Such hypothetical mechanisms could not be tested in the present because they were so large and had only operated in the past! Hutton made several important conceptual advances: insisting on looking for mechanisms that were measurable in the present, noting that there was no available evidence (at that stage) for when the earth was actually formed, realizing a long timescale must exist, and showing that (given such extremely long time periods) relatively small forces could lead to major changes on the surface of the earth. One of his statements concluding his first chapter was '*we find no vestige of a beginning,—no prospect of an end*'. His main conclusion had been published earlier in 1785. He saw no existing evidence (at that time) for the age of the earth, but it had existed for a very long time, geologically speaking (see Dean 1992).

Perhaps in Hutton's time, there was not enough background knowledge to really challenge the popular theories that 'explained' past history (geology and biology) in terms of mechanisms that were not testable. One of his statements emphasized this; he argued for more limited hypotheses that were judged by their consistency with the evidence: '*and there may be many causes of which we are as yet ignorant*'. Hutton is often remembered today as discovering 'unconformities', where there had been a long period of net erosion between the time periods that two strata had been laid down. For example, the earliest stratum might be now vertical and the later (upper) stratum horizontal. As the following extract (from Chapter 9, sect. 1, p. 547) notes, he had been searching for the phenomenon before he found an example in southern Scotland. '*When one day, walking in the beautiful valley above the town of Jedburgh, I was surprised with the appearance of vertical strata in the bed of the river, where I was certain that the banks were composed of horizontal strata. I was soon satisfied with regard to this phenomenon, and rejoiced at my good fortune in stumbling upon an object so interesting to the natural history of the earth, and which I had been long looking for in vain.*' The main aspect from this extract is that Hutton had predicted a long timescale and then searched for the evidence (for or against).

He had similarly found evidence for geological rates in old Roman roads going across southern Scotland. The rate of erosion was relatively slow—and he could still find the old pits that had been used to build the roads nearly 2,000 years previously. Similarly, he conjectured and then tested his suggestion as follows: '*I have always conjectured that the waters of Giezer* [Iceland] *must be impregnated with flinty material by means of an alkaline substance, and so expressed my opinion . . .*' (p. 432). He had analyzed material bought back from Iceland by Joseph Banks; then when Mr Stanley went, he arranged for him bring back more samples of rocks and water—he got both siliceous and calcareous deposits from the water. Measurements were by Dr Black. '*My conjecture has thus been verified.*' It was good testable science, though perhaps still a little lacking in modern knowledge.

However, thirty-five years later (in the 1830s), the increase in knowledge was sufficient for Charles Lyell to eventually convince more geologists that a long timescale existed (many millions of years). The early concepts of species had not considered a timescale of many millions of years. His *Principles of Geology* was published in three volumes (1830–1833). Lyell is a central figure for Darwin's thinking (see later) and his scientific approach. Lyell's book was critical both in setting both

> a much longer timescale (many millions of years) and standards of scientific explanation for past events.

The book's subtitle is important, and it should be read carefully: *Being an Enquiry How Far the Former Changes of the Earth's Surface Are Referable to Causes Now in Operation*. It indicates that Lyell strove to interpret past events, not by imaginary mechanisms that only occurred in the distant past, but by the same types of mechanisms that can occur in the present—probably *actualism* rather than uniformitarianism.

The difference between uniformitarianism and actualism is that uniformitarianism (which was coined by one of Lyell's opponents, William Whewell (1794–1866)) is ambiguous and generally assumes constant rates, whereas actualism assumes the mechanisms are the same even though the rates can vary. Whewell's well-known statement comes from his review of Lyell's book. Whewell (who was influential in his own right) was one of Lyell's (and later Darwin's) opponents. The quote is important in that it has been surprisingly influential: '*These two opinions will probably for some time divide the geological world into two sects, which may perhaps be designated as the "Uniformitarians" and the "Catastrophists". The latter has undoubtedly been of late the prevalent doctrine, . . . Indeed, we think it ought to be so.*' Lyell responded in his third volume (which was published over a year after Whewell's review was published). His response was '*We regret, however, to find that the bearing of our arguments in the first volume has been misunderstood in a similar manner, for we have been charged with endeavouring to establish the proposition, that "the existing causes of change have operated with absolute uniformity from all eternity." It is the more necessary to notice this misrepresentation of our views, as it has proceeded from a friendly critic whose theoretical opinions coincide in general with our own, but who has, in this instance, strangely misconceived the scope of our argument'* (vol. III, p. 126). Despite this clarification by Lyell, it does appear that Whewell's interpretation is the 'standard' one, even today. Surely we cannot be that stupid so as to keep an error for almost 200 years! But it does look like it.

Anyway, humans do appear to like 'binary choices' and so often (usually?) seem to want this type of forced choice, but we really should resist being forced into binary boxes in this way. It is probably more helpful to group the different ideas into at least four major viewpoints:

> catastrophism (many authors, for example, Cuvier),
> orthogenesis (for example, Lamarck),
> actualism (Hutton, Lyell, Darwin), and
> uniformitarianism (many authors).

Things did not have to change at a 'uniform' rate; rates could vary, but the mechanisms would be the same. Actualism, uniformitarianism, and orthogenesis could all be grouped as 'gradualism' in that they emphasize smaller and slower changes and a longer timescale. Similarly, catastrophism and orthogenesis could be grouped because they do not emphasize testable mechanisms; they really just assert that 'something is true'.

In the following analysis (from Penny 2009a), we divide Whewell's concept of uniformitarianism into five sections. Lyell would have accepted three of them (indicated by a √) and rejected two of them (indicated by an X). Note that at that time, Lyell rejected evolution.

> √ 1. The laws of nature are 'uniform'. In his frontispiece to his *Principles*, Lyell uses an extract from John Playfair ((1748–1819) a biographer and interpreter of James Hutton): '*The economy of Nature has been uniform, and her laws are the only things that have resisted the general movement. The rivers and the rocks, the seas and the continents have been changed*

in all their parts; but the laws which direct those changes, and the rules to which they are subject, have remained invariably the same.'

√ 2. These fundamental physical laws led to variable rates of change of geological processes, including uplift, erosion, volcanic activity, climate change, etc.

√ 3. Regional and larger catastrophes are part of the system, but they operate by known (or knowable) mechanisms, including comet impacts.

X 4. There is no 'progression' of plant and animal forms through time. In the early 1830s, Lyell rejected any form of 'evolution'—whether natural or guided. Later (possibly under Darwin's influence?) he accepted evolution.

X 5. Large catastrophes by unknown (or unknowable) mechanisms are necessary to account for past changes.

Actualism is therefore a more accurate description of Hutton's, Lyell's, and Darwin's ideas than the term 'uniformitarianism' (as defined by Whewell). A major area of difference between the two is that actualism allows both variable rates and a few 'catastrophes'—provided they occurred by mechanisms that could still be studied in the present. Lyell gave at least four major examples of catastrophes that could happen, including:

> a massive outflow from the Great Lakes of North America (if the Great Lakes were breached),
>
> the Mediterranean drying up if the seabed at the Strait of Gibraltar was raised sufficiently to cut off the inflow from the Atlantic Ocean,
>
> a massive inflow if the Mediterranean was subsequently reconnected to the Atlantic by lowering the Strait of Gibraltar, and
>
> flooding much of Central Asia if the land between the Caspian and Black Seas was lowered (it was known that some of the inland areas were below sea level).

We will see later that some impacts were also accepted, but their effect could still be studied in the present. The ideas are fairly similar, but often the term 'uniformitarianism' is (usually) taken to imply constant rates of processes—not just the same laws.

Lyell did not insist on a theory explaining everything ('life, the universe, and everything'). For example, Lyell refused to 'speculate' on the origin of the earth because he did not have evidence on which to base a scientific theory of the origin of the earth—it was outside his sphere of interest. As mentioned earlier, from the late seventeenth to early nineteenth century, there was considerable antagonism in science to 'speculating' on the origins of things. A good early example is one we have already mentioned: Isaac Newton (1687) similarly refused to 'speculate' on the origin of the solar system. Similarly, Lyell (1830) refused to 'speculate' on the origin of species. Lyell's early view at that time was that once species were formed ('created'), they were subject to normal physical and biological principles, even to the extent of becoming extinct. Scientific laws are made, and then the system was left to its own devices. But later scientists (such as Darwin) were able to put forward testable scientific hypotheses for the origin and evolution of species.

Thus a major breakthrough resulted through both improvements in geology recognizing the long timescale of the earth, together with a mechanistic process of explaining past events by 'causes still in operation' or 'physical laws still operational'. The long timescale meant that relatively small forces (such as erosion) acting over long periods of time could lead to quite major geological changes. Similarly, it allowed ample time for evolution to occur. Charles Darwin started his professional career as a geologist (see later) and then introduced this more rigorous scientific approach into historical questions in biology. The Hutton and Lyell tradition in geology (that Darwin adopted) was mechanistic in the sense that former changes should be explained by causes (mechanisms) that were still (in principal) operational, including catastrophes by known mechanisms.

An interesting feature of Lyell's *Principles of Geology* is that there is a large amount of 'biology' in it, especially in Volume 2. Lyell was interested in all factors, physical and biological, that affected geological processes (including erosion and sedimentation). At the time of Lyell's writing (1830s), geology covered many historical aspects of biology, including biogeography—his second volume is perhaps the best ecology text in English for the time. It is still very interesting reading and introduces 'modern' theories such as the Red Queen hypothesis (having to continually run fast just to stay in the same place). Lyell's biological knowledge was aided by a visit across the Alps from Italy to Switzerland (in wintertime) to a visit a Swiss plant ecologist (Augustin de Candolle (1778–1841)). The visit occurred after Lyell had met one of De Candolle's former students and realized that De Candolle had some interesting information. It is probably by this route (De Candolle via Lyell) that Darwin learned many important ecological ideas, including competition. We will see in a minute or two that De Candolle considered competition virtually universal in biology. Many early developments in ecology occurred in southern France and in Switzerland and were virtually unknown by English-speaking ecologists. Fortunately the ideas got into biology through the early geologists, both Hutton and Lyell.

The rest of Lyell's second volume discusses biogeography, mechanisms of dispersal of plants and animals, the potential for increase in population numbers, limitation of resources, the regulation of population numbers, estimates of the rates at which species became extinct, etc. Lyell was well aware of competition for resources between plants and quotes from Augustin de Candolle's book phrases such as the following:

> *'The most fertile variety would always, in the end, prevail over the more sterile'* (vol. II, p. 34).
> *'Unhealthy plants are the first which are cut off by causes prejudicial to the species, being usually stifled by more vigorous individuals of their own kind. . . . In the universal struggle for existence, the right of the strongest eventually prevails'* (pp. 55–56).

'*All plants of a given country . . . are at war, one with another. The first which establish themselves by chance in a particular spot, tend . . . to exclude other species, the greater choke the smaller, the longest livers . . . the more prolific. . . . "In this continual strife, it is not always the resources of the plant itself . . . Its success depends, in a great measure, on the number of its foes and allies among the animals and plants inhabiting the same region*' (vol. II, p. 131).

Until around the end of the nineteenth century, geology included many aspects of environmental and historical biology as well as physical geography. Lyell also frequently quotes Prichard's *Researches into the Physical History of Man* (1813), which also was an early attempt to give a naturalistic approach to biological science.

A contrast with the approach of Hutton and Lyell is exemplified by French Georges Cuvier ((1769–1832) see Cuvier 1817), who applied his knowledge of anatomy to fossils and thus could estimate how the fossil animals would have lived. By comparing fossils from different locations, he could identify equivalent geological strata; this allowed the recognition of strata of the same age, but at different locations. Thus he founded stratigraphy in geology (along with William Smith in England; see Winchester 2001). However, Cuvier assumed that species were constant within strata, but that there were a succession of creations and destructions (catastrophes). Perhaps his apparently rigorous standards meant that he was more limited by the knowledge of the time, particularly by the supposed short geological timescale.

We have already mentioned an example of his reasoning in the argument that skeletons of cats, dogs, and other animals (including birds) from the Egyptian pyramids (from about 2000 BCE) were virtually identical to modern forms; consequently, these species had not changed in about 4,000 years (*c.*2000 BCE to AD 1800). Therefore, Cuvier concluded that evolution had not occurred because species had not changed! He assumed catastrophes turned over a fair proportion of the earth's surface by mechanisms that could not possibly be studied today. His book *Essay on the Theory of the Earth* (1817) was a disagreement with the *Theory of the Earth* by James Hutton (1795), whom we have already seen argued for both very long (and unknowable?) timescales and explained the past strictly in terms of mechanisms that could be studied in the present.

The following extracts come from Cuvier's *Essay on the Theory of the Earth*, translated in 1817. Cuvier considered that the discovery of frozen mammoth carcasses in the Arctic must 'remove for all time' any thought of explaining climate changes (such as the onset of Ice Ages).

> *These repeated irruptions and retreats of the sea have neither been slow nor gradual; most of the catastrophes which have occasioned them have been sudden; and this is easily proved, especially in regard to the last of them . . . the carcasses of some large quadrupeds which the ice had arrested, and which are preserved even to the present day with their skin, their hair, and their flesh. If they had not been frozen as soon as they were killed they must have quickly been decomposed by putrefaction. . . . Life, therefore, has been often disturbed on this earth by terrible events—calamities which at their commencement, have perhaps moved and overturned to a great depth the entire outer crust of the globe, but which, since these first commotions, have uniformly acted at a less depth and less generally. [Footnote]*The two most remarkable phenomena of this kind, and which for ever banish all idea of a slow and gradual revolution, are the rhinoceros discovered in 1771 . . . and the elephant found . . . . The last retained its flesh and skin . . . was still in such high preservation, that it was eaten by dogs* (pp. 15–16).

Pity no one told Ötzi (the Ice Man), who got trapped in the ice about 5,000 years ago and was discovered only in 1991. Ötzi had not degraded away (he had been frozen the whole time)!

Cuvier certainly accepted that the norm was to explain past historical/political events in terms of human behavior or instincts.

As it has been long considered possible to explain the more ancient revolutions on its surface by means of these still existing causes; in the same manner as it is found easy to explain past events in political history, by an acquaintance with the passions and intrigues of the present day. But we shall presently see that unfortunately this is not the case in physical history; the thread of operation is here broken, the march of nature is changed, and none of the agents that she now employs were sufficient for the production of her ancient works (p. 24).

So his interpretations of natural events were very different whether they were political or whether they concerned natural phenomena.

His concluding reflections are illustrated below:

I am of the opinion, then, with M. Delac and M. Dolomieu,— That, if there is any circumstance thoroughly established in geology, it is, that the crust of our globe has been subjected to a great and sudden revolution which cannot be dated much further back than five or six thousand years ago; that this revolution had buried all the countries which were before inhabited by men and by their animals . . . (p. 171).

Thus Cuvier was an outstanding scientist in that he helped found geological stratigraphy but was still limited by his acceptance of a shorter timescale. His ideas live on; in Chapter 6, we consider the loss of the dinosaurs by a similar proposed mechanism.

Perhaps the main point of what we learn from Cuvier's quotes is that the long geological timescale and that science uses 'known mechanisms' were not obvious to many people at the beginning of the nineteenth century. Nevertheless, both are of fundamental importance for evolution as we know it. Although we now take a long geological timescale for granted, earlier scientists would have had to fit all of evolution into a few thousand, or tens of thousands of, years. This meant that it was apparently impossible to explain any form of evolution by means that we would now consider standard. Much later we got independent checks on the ancientness of the earth from radioactive dating, but the methods did not exist at these earlier times. It did need a timescale of millions of years for evolution to occur. Good biologists, such as Cuvier, could not/ did not see the very long timescale. We should be very grateful to those researchers who did move our knowledge forward, and we certainly now accept a geological timescale of many billions of years. This is the first main point of the chapter.

The Young Charles Darwin as a Geologist

Many biologists do not seem to accept, or to like, the information that Darwin's first profession was as a geologist. After dropping out of medical school in Edinburgh (he didn't like the sight of blood), Darwin then took a general degree at Cambridge, probably intending to become a village parson. But on completing that degree, he left on an around-the-world voyage on the *Beagle* (1831–1836) convinced of finding evidence for the fixity of species. The *Beagle* was under the command of Captain (later Admiral) Fitzroy (who was also later the governor of New Zealand). Darwin records in his autobiography that as an undergraduate at Cambridge, he had been strongly influenced by William Paley's natural theology and the argument from design (see earlier); he found its logic to be very powerful. Finding a complex object such as a watch that 'exists for a purpose' leads, as illustrated in the earlier, to the conclusion that there exists (or existed) a 'watchmaker' because the watch was formed for a purpose. Similarly, finding intricate and complex organisms well suited to their environment and mode of life appeared to imply both a Creator and a purpose to the parts of the organism. Acceptance of natural theology was Darwin's starting point at the beginning of the voyage of the *Beagle*.

An important figure was John Stevens Henslow (1796–1861), professor of geology (1822–1828 and professor of botany (1825–1861), both at

Cambridge University. He was originally offered the position as the naturalist on the voyage of the *Beagle* but declined (on the advice of his wife) and suggested Charles Darwin instead. Before the *Beagle* left England, Henslow gave Darwin a copy of the first volume of Charles Lyell's *Principles of Geology*:

> *When I was starting on the voyage of the Beagle, the sagacious Henslow, who, like other geologists believed at that time of successive cataclysms, advised me to get and study the first volume of the Principles, which had then just been published, but to on no account accept the views therein advocated. . . . I am proud to remember the first place, St. Iago in the Cape Verde Archipelago, where I geologized, convinced me of the infinite superiority of Lyell's views over those advocated in other work known to me* (Darwin 1969: 101).

So the warning did not work! During the voyage of the *Beagle*, Darwin was strongly influenced by Lyell's logic, especially whether past geological events were caused by mechanisms similar to those currently occurring—and that could be studied 'in the present'.

During the voyage of the *Beagle*, Darwin collected widely and carried out geological fieldwork ashore. He became convinced of the superiority of Lyell's mechanistic approach to geology and later the two became good friends. Darwin wrote three books on the geological findings of the voyage (*Geological Observations on South America, Structure and Distribution of Coral Reefs*, and *Geology of Volcanic Islands*). He says of his book *Coral Islands*, '*No other work of mine was begun in so deductive spirit as this; for the whole theory was thought out on the west coast of S. America before I had seen a true coral reef. I had therefore only to verify and extend my views by careful observation of living reefs. But it should be observed that I had during the two previous years been incessantly attending to the effects on the shores of S. America of the intermittent elevation of the land, together with denudation and the deposition of sediment* (1969: p. 98).

The following six points demonstrate his early professional interest in geology.

1. He wrote the three books on the geology of the voyage (*Geology of South America, Coral Islands, Geology of Volcanic Islands*), but he only edited the zoology publications and was not involved with botany publications (his samples were 'lost' for well over 100 years).

2. His letters during the voyage of the *Beagle* show confidence on geological subjects but also show his concern about the (supposed) inadequacy of his biological collecting.

3. Some of his geological letters to colleagues had already been published before his return from the voyage of the *Beagle*.

4. On his return from the voyage, he joined the Geological Society of London (rather than the Linnean Society or the Zoological Society).

5. He presented papers to the Geological Society, published in its geological journal, was elected to its council, and became its foreign secretary.

6. Soon after the return of the *Beagle*, he became a close associate with the leading geologist, Charles Lyell, and was a frequent visitor to his London home.

Charles Darwin's geological career is now written up (Herbert 2005), and by the end of the voyage, he was considered a leading geologist. His talks to the Geological Society showing that Lyell's ideas applied equally well to the southern hemisphere and the other side of the globe (South America in particular) helped convince other geologists that Lyell's approach and reasoning were correct. Figure 2.1 illustrates how early in his career his published papers were mainly in geology and only later on biological topics.

Darwin's papers by subject area.

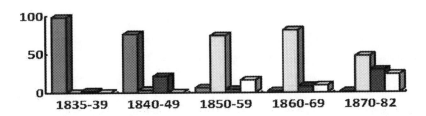

Fig. 2.1. Darwin's publications by subject area and time. For each time period the number of pages of scientific papers (excluding books) is counted by subject (Barrett 1977) and then expressed as a percentage of total pages published. Time periods are based around 1859 when the Origin was published. The figures are for Geology, Botany, Zoology and General (includes psychology and child development as well as evolutionary theory). The figure illustrates both why the young Darwin is considered a geologist, and the general nature of his later interests.

By the end of the 1830s, Darwin had extended Lyell's reasoning into biology and the origin of species, though it was many years before he published his conclusions. James Hutton had earlier accepted that 'known processes' might lead to varieties that could arise through the processes operating in the present (but apparently not to new species). Lyell had also partially applied this reasoning to biology—once they were created, species were subject to normal scientific laws, even to going extinct. Lyell had refused to 'speculate' on the origin of species, but once species were created, they were subject to normal physical and biological laws. Darwin went further and looked for 'causes now in operation' to explain changes in and between species—in other words, mechanisms that could be studied in the laboratory or in the field; untestable mechanisms were to be avoided because such ideas were metaphysical (such ideas may eventually be found to be correct, but they were untestable at present). So he went another step; he refused to 'speculate' on any scientific mechanism for the origin of life, but once life occurred, then biological and physical laws were apparently sufficient to account for the origin of new species.

Biologists (for some unknown reason) do not seem to like to think of Darwin's early professional career in geology, but it was important for his mechanistic approach. So this is the second theme of the chapter. The next section returns to biology and considers some of the many early ideas of the relationships of species.

Early Ideas on the Patterns or the Relationship between Species

In this section, we study some of the relationships between species that were proposed between 1700 and before about 1850. The most common pattern relating living organisms (particularly animals; people studying plants were maybe not so convinced) was the Great Chain of Being (Figure 2.2a)—a linear hierarchy of species. Humans were (of course) at the top of the chain, but just to keep us humble, there were various forms of angels and archangels above us. And just to please the females, males considered themselves superior. And of course, Parisian males considered themselves at the top, but Berlin males knew they were at the top. Then there were the English males, who considered London males at the top of the scale! However, females knew they were really at the top of the chain!

Care needs to be taken in examining the diagrams in Figure 2.2. It cannot be claimed that any figure represents exactly what any particular scientist thought. In most cases, we do not know their precise thoughts; they usually did not put their ideas in diagrammatic form, and their ideas usually changed as they studied additional plants and animals. Rather, what is intended is the reverse. What are the possible relationships where at least one biologist at one time expressed ideas similar to those shown in the diagram?

Initially the Great Chain was usually considered as static—that is, species were created in their present form and in their present place on the Great Chain. However, a few authors considered that 'species' (or perhaps better populations or 'monads') were either degenerating down the Great Chain or ascending up it. It was assumed that there should be a more or less continuous series of intermediate forms on the Great Chain. Consequently, finding 'missing links' was evidence that the Great Chain was indeed the Creator's plan for the universe. For example, the discovery of chimpanzees that fitted between humans and monkeys was a cause for celebration in the late seventeenth century; indeed the chimpanzee's first scientific name was *Homo sylvestris*— human of the forest (Tyson 1699). The proposed idea of a continuous series of intermediate forms fitted with the idea of continuity in nature. There should be no gaps in nature: *natura non facit saltum* (nature does not make jumps). As an aside, perhaps, we think that they got it right in 1699. *Homo sylvestris* (human of the forest) is an appropriate name for chimps compared with *troglodytes*—cave dwellers. And the genus *Equus* (horse and zebras) diverged about 5–6 million years ago, about the same time as humans and chimps. So why are we not all three species (humans, chimps, and gorillas) of *Homo* (Watson et al. 2001)? (The name *Homo sylvestris* for chimpanzees was before Linnaeus, so it doesn't count officially!)

We have already met Buffon (Georges-Louis Leclerc, Comte de Buffon), who was one of the most influential naturalists in the eighteenth century, and he put forward an alternative view to the Great Chain. Among his prolific output, he considered the possibility that some closely related species (say, the great-cat carnivores—lion, tigers, leopards, etc.) could be degenerated from an original 'ideal' form of great cat. Modern forms had degenerated from this ideal form, and this possibility is shown in Figure 2.2b. For mammals, he suggested about forty original forms of mammals, perhaps more or less equivalent to families in modern classification. So this allows another form of relationship, the special creation of a few forms and 'degeneration' into modern species. This allows a further modification to Ray's original concept of species.

Several early scientists did suggest an evolutionary origin of species. Erasmus Darwin (physician, scientist, poet, inventor, and grandfather of Charles Darwin) developed an evolutionary theory in the late eighteenth century. Perhaps it was more a progenitor of Lamarck's theory rather than of Charles Darwin's theory—it was not mechanistic in the sense of specifying how species changed with time. Among many other things, he did turn Linnaeus's classification of plants into poetry (Darwin 1791; King-Hele 1999)!

Another early evolutionary theory was proposed in France by Jean-Baptiste Lamarck (1744–1829). His books of 1802 and 1809 in particular proposed a genuine evolutionary theory (Lamarck 1809; Burkhardt 1977). He was convinced of evolution by gradation between species, though his proposed mechanism (necessarily) was rather vague:

1. need or striving of animal (not necessarily the 'will')
2. inheritance of acquired characters (this is the Old Testament view)

Lamarck appeared to assume that a succession of new life forms ('monads') arose spontaneously and gradually evolved up the Great Chain into 'higher' forms. Note that he apparently did not allow extinctions—species automatically adapted to their needs. A problem with his theory was that the mechanism was not testable at that time—it did not lead to experiments or draw support from earlier experiments. Perhaps he was the last of the eighteenth-century scientists for whom a theory was judged on whether it could explain everything rather than on how well it was corroborated or supported. Lamarck had other theories that 'explained' all of chemistry or all of meteorology, again by appealing to forces and mechanisms that were untestable. Figure 2.2c gives a likely illustration of Lamarck's ideas of evolution; new 'life forms' or 'monads' are related indirectly—that is, their similarity did not result because they shared a common ancestor. Species continually arose and kept on 'improving themselves'.

The form of the improvement would depend on the environment in which they were living. Some splitting of lineages was possible, but similar-looking species (such as lions and tigers) need *not* be closely related. Instead, they could be separate 'monads' that had formed spontaneously but always found themselves in similar environments. His theory was certainly more sophisticated than just 'inheritance of acquired characters', which is a modern parody of Lamarck. At that time, everybody accepted at least some inheritance of acquired characters; it was certainly not unique to Lamarck. However, there were several other evolutionary theories, including those of the German philosopher Johann Wolfgang Goethe and Erasmus Darwin, Charles Darwin's own grandfather.

Similarly, we have already discussed some of ideas of Georges Cuvier. He emphasized 'conditions of existence', a static creationist view of species; every item of the organism was perfectly designed for the environment in which it was destined to live—a highly functional view of organisms. This is represented as Figure 2.2d, no overall pattern of relationships. He was the main opponent of the evolutionary ideas first of Lamarck's and later of Étienne Geoffroy Saint-Hilaire's. Although Cuvier was 'non-evolutionary', he did introduce some more rigorous standards into some other natural history aspects of biology, and he did appear to have respect for Lamarck.

Another important figure was the French naturalist Saint-Hilaire (1772–1844). At one time, he emphasized 'unity of type' and looked for patterns among organisms rather than considering each species separately—searching for an archetype or basic plan or 'idea'. He assumed some form of evolution but was even vaguer than Lamarck on the possible mechanisms involved. He answered Cuvier's argument (mentioned earlier) by pointing out that the climate of Egypt had not changed over the last 4,000 years; therefore the species had not changed! Geoffroy suggested relationships between widely different groups (even vertebrates and invertebrates) based on apparent morphological similarities. He was important in developing ideas of common pattern in animals but is regarded by many as 'nonscientific' (perhaps he should be called 'pre-scientific') because there was no way at that time of testing the ideas. Modern developmental genetics is vindicating aspects of his relationships of form between invertebrates and vertebrates. For example, he suggested that some ventral features of vertebrate organs were homologous to dorsal organs in invertebrates. There was a major debate in Paris between Cuvier and Geoffroy (Appel 1987).

Also important was Richard Owen (1804–1892) in England. He was similar to Geoffroy in being an outstanding anatomist, but he assumed archetypes were the basic plan of the Creator. He accepted the concept of *archetype*, but the archetype had perhaps never existed as a real organism but was only a plan or 'idea' of the Creator. Every species created was then modified from this plan to the particular conditions in which they existed. This was a common viewpoint and is indicated as a series of 'star' trees—each species is created independently by independent modifications to the archetype. A useful analogy is an architect who has a plan of an 'ideal' house, but any particular house had to be modified to the conditions of its own environment—slope, shade, climate, etc. We can represent this as Figure 2.2b, but in Buffon's view, the archetype had actually existed. Louis Agassiz (Switzerland and United States), who popularized the important concept of an earlier Ice Age, appears to have had similar ideas to Richard Owen in this respect.

Other patterns for the relationship between species were considered in the first half of the nineteenth century and are illustrated as Figures 2.2e–h. Thomas Huxley briefly considered a two-dimensional graph, a bit like Figure 2.2e; each species would have four closest neighbors. This is a bit like Mendeleev's table of the chemical elements. (Earlier, Linnaeus wondered whether the families of flowering plants might fit a two-dimensional map, much like countries in an atlas. Linnaeus also suggested that some plant species may have arisen from hybridization; this could also give a two-dimensional system.)

Figure 2.2f shows the quinarian (five) circular or osculating system of William MacLeay (England, later in Australia), where there are always five groups arranged in a circle. Where the circles touch (osculate or kiss), there are members of the group with some features found also in the adjacent group. Each of the groups is subdividable into five more. This was suggested to be the plan of the Creator. William Swainson (also in England, later in New Zealand) was one of the few who followed the system, and he applied it to bird classification.

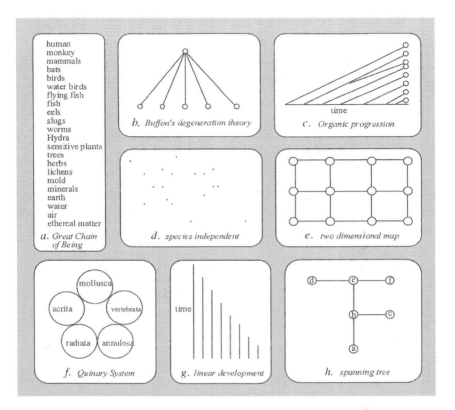

human
monkey
mammals
bats
birds
water birds
flying fish
fish
eels
slugs
worms
Hydra
sensitive plants
trees
herbs
lichens
mold
minerals
earth
water
air
ethereal matter

a. Great Chain of Being

b. Buffon's degeneration theory

c. Organic progression

d. species independent

e. two dimensional map

mollusca
acrita
vertebrata
radiata
annulosa

f. Quinary System

g. linear development

h. spanning tree

Fig. 2.2. Relationships between species considered by early biologists: **(a)** the Great Chain of Being, favored particularly by zoologists. Most authors considered it static. Others imagined species ascending the chain with time. Some thought species degenerated with time. **(b)** A number of species 'degenerated' from one original (for example, giant cats from one original 'perfect' form), but the figure could represent several species created from a perfect archetype (idea) with each species modified for local conditions. **(c)** A representation of the idea of continued spontaneous generation of new 'monads', which then ascended a form of the Great Chain of Being (shown with limited speciation). **(d)** A form of special creation where each species was designed for its environment without any overall pattern. **(e)** A two-dimensional map with species or groups of species being either in a regular arrangement or occupying defined regions. **(f)** The quinary system with five 'osculating' circles that were repeated at each level of classification; intermediate forms occurred at each intersection (osculation). **(g)** A linear development with species arising at the same time but ascending to 'higher' forms at different rates. **(h)** A spanning tree that links existing species. Species 'b' could be derived from species 'a' (which remained unaltered). Species 'c' could be derived from 'b' and so forth. The figure shows that many forms of relationship were possible; it was a major achievement of biology to decide on a tree by inserting ancestral states (see later). A tree is an essential part of the model biologists now seek to reconstruct. Early biologists certainly had no a priori reason to reconstruct a tree (based on Penny et al. 2003).

The next scheme, Figure 2.2g, may represent some of the ideas of Robert Chambers, who in 1844 published (anonymously) an early book on evolution: *Vestiges of the Natural History of Creation*. In his version, different species ascended the Great Chain at different rates. Finally there is the idea that one species could be created (or modified) from another species (see Figure 2.2h). Because the direction of change is not specified, the tree is shown as an undirected (unrooted) tree. This differs from Figure 2.2b in that the ancestral species can still exist.

Overall, the main point to remember from this section is that there were many attempts to find the pattern of relationships between species, and many alternatives are discussed by Bowler (2003). The most productive scheme—that of Darwin's evolutionary tree (descent with modification)—was neither obvious nor easy to find. A real difficulty was trying to fit everything into a short timescale, so it really helped when geologists (and later physicists) found that the world had existed for many, many millions of years.

Thus it is often helpful to look at the overall picture to see the general context of a subject matter. Where did the new ideas come from, and how did they relate to the existing ideas? The following summarizes five key points of these first two chapters. In the next chapter, we will consider Darwin's theory in detail. But in the meantime, we should be aware that there were many ideas for the relationship between taxa/species, and we should also be aware that peoples' ideas do change with time.

Five Points

So we need to briefly summarize these first two chapters. Perhaps the most important thing is to accept that the increase in scientific knowledge gives us both new opportunities and new responsibilities (such as knowledge about global warming). Without our knowledge to understand the past, we are not able to predict the future and adjust accordingly—science gives us this opportunity. We must be careful not to put unfair judgments on earlier peoples, who acted according to their knowledge at the time. But neither should we distort the past in order to claim credit in the present; humans love to blame other people. But given that many indigenous people around the world (including Europeans) assumed that forms or kinds of life continued to regenerate from the soil, is it helpful to 'blame' earlier people if they, perhaps by overhunting, caused the extinction of species (such as the Irish elk in Europe or moa in New Zealand)? We need a fairer answer.

The development of the concept of species is an important step in the rise of modern science—an important advance at the time. Perhaps six important aspects should be made:

> There is no continued spontaneous generation.
> Members of a species are an interbreeding group.
> Species have some permanence through time.
> It was essential to have a long geological timescale.
> The young Darwin started out as a geologist.
> Many suggestions were made for the relationship between taxa.

It was a major advance in the 1600s to suggest that species could have a permanence for 6,000 years or so. Today we go even further and see that some species (e.g. the tuatara) have existed for many millions of years. It was the discovery during the nineteenth century of a much longer timescale for the solar system that allowed the important step to the modern understanding. The early species concept (late seventeenth century) assumed that each species had an 'unchangeable essence' (as expected from Plato's philosophy). We do not accept this today.

A major point was the development (in geology and then in physics) of a long timescale for events in the history of the earth to occur. The denial of both transformations between living and non-living forms and of continuing spontaneous generation, together with the existence of interbreeding and long-lasting species, meant that there was a problem in the origin and relationship between 'species'. For a relationship between species, the Great Chain of Being was originally perhaps the most popular description; many others were considered. Over 150 years elapsed from the development of the species concept until Darwin suggested the pattern of a tree with ancestor/descendant relationships (though we would now allow networks). There were other theories that had species changing with time, but they all lacked a mechanism where the components could be tested independently. In addition, the short timescale appeared to preclude slow changes leading to new species. It seems to be important that Darwin started off his professional career as a geologist and was quickly convinced by the very mechanistic approach of Charles Lyell.

The early history of the species concept and of the relationships between species is interesting in its own right. However, it is important in illustrating that science is an ongoing process; although we study it at one point it time, science is still unfolding. Sometimes by examining the origin of a concept, we can see ways of improving it. There is no reason whatsoever to assume we are at the endpoint of scientific understanding; people in the future will look back and smile at our 'quaint' ideas. Dammit, we just don't know which of our current ideas are the ones that will be considered 'quaint' in the future. But later we will look at a couple of possible ideas; the trouble with science is that it is done by humans!

CHAPTER 3

The Structure of Darwin's Theory: Microevolution, Macroevolution, and Microevolution as Sufficient for Evolution

In this chapter, I will discuss an analysis of one way to analyze the many components of Charles Darwin's theory. There is a tendency of humans to want to simplify ideas into 'yes' or 'no' answers, but life is more interesting than that. Similarly, there is a tendency for a few individuals to say that they favor some (usually unspecified) 'non-Darwinian' mechanism. But if we specify some twenty factors that contribute to Darwinian evolution, we see that some people just do not say what aspect they accept or don't accept. Within this framework of 20 points, we will also consider three main hypotheses about his mechanism—namely microevolution, macroevolution, and that microevolutionary processes are both necessary and sufficient for macroevolution.

We have discussed that Charles Darwin started his professional career as a geologist and was very quickly convinced of Lyell's approach to geology. However, in those days of more generality in science, he had sufficient knowledge of biology to interpret the consequences of what he saw. During the voyage of the *Beagle*, Darwin was influenced by at least four biological observations:

1. Variation within species along the coasts of South America
2. Finding fossil remains of extinct species (but some had living relatives)
3. Variation between islands in the Galapagos group
4. The relation of Galapagos species to those of mainland South America

This last observation, for example, argues against Galapagos species being independently created. Rather, it appeared their ancestors had arrived by natural means from the nearest land and then had become modified to the local conditions. Consider an alternative: if you (or the Great Gardener in the Sky) were going to find plants and animals for a new volcanic island (and scientists accepted that the Galapagos Islands were geologically young volcanic islands), you could select them from anywhere on earth that had a similar climate, not just from adjacent regions (that may not have quite the same climate and soil types). In contrast, in nature it seemed the species for the new islands came from the adjacent land and had adapted to local conditions. Consequently, it appeared simpler if species had arrived by natural means, not by the deliberate planning of a Great Gardener in the Sky. So biogeography was important in leading to the conclusion that evolution had occurred and that there was continuity between present and earlier species. So even though Darwin was still a geologist at that time, he knew enough about 'nature' to know some of the important questions.

Fig 3.1 - 1837 figure

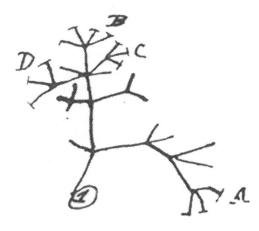

By 1837 (after the voyage), he was convinced that evolution had occurred, and he recorded a branching diagram in his first notebook of 'transmutation of species' (Figure 3.1, de Beer 1960: p. 46), a reduced version to that shown later in *Origin*. This is part of his overall theory 'the theory of descent', and the conclusion implies that relatedness of species is a consequence of descent from common ancestors. In the mid-1830s, he could not think of a realistic mechanism to explain the divergences and especially the adaptation of each species to its environment.

While looking for a possible mechanism, he read very widely. He was influenced by the French philosopher Auguste Comte (1798–1857, e.g. 1835) on what was expected of a good scientific theory. Comte proposed three stages of scientific enquiry: (1) theological, (2) metaphysical (or teleological), and (3) fully 'scientific'. This classification maybe too simple, but it is useful, especially for understanding Darwin's theory compared with earlier ideas. In a letter to Charles Lyell in 1838, he writes that Comte has '*some fine sentences about the very essence of science being prediction*' (Burkhardt and Smith 1986: 104).

'Theological' theories in science are perhaps self-evident—the creation of new species and moving the stars across the sky by angels (or horses) are examples. 'Metaphysical' is a little more complex. A good example is the ancient Greek explanation of a stone being released from your hand and falling to the earth. At that time, there were assumed to be four 'elements' (earth, air, fire, and water); stone was mostly earth, and earth 'knew' its place was at the center of the universe, and so the stone therefore moved towards the ground to its natural place. (This might reinforce the idea that the earth was the center of the universe.) But how do you measure the property of the stone 'knowing' it is 'mostly earth'? How do you measure its ability to 'sense' in which direction is the ground? Such explanations are not really independently testable.

An equivalent example in biology are many forms of 'orthogenesis' as an explanation for a mechanism of evolution—somehow living matter has the ability to change, slowing over long periods of time to a 'better' or 'improved' state; but this would be a 'property' of living systems that can't be measured. As we said earlier, it's a pity no one told the tuatara (*Sphenodon*) or the horseshoe crab (*Limulus*) or the maidenhair tree (*Ginkgo*) about the theory; those three groups might have just kept evolving! Today the three groups are often called 'living fossils' precisely because they do appear to have remained morphologically static for a 10 or 100 million years or so; they have not kept evolving.

To Auguste Comte, a scientific theory must use ideas or properties which are independently measurable. This would lead to predictions that were testable. Perhaps Comte's three 'stages' may no longer be popular as a description of scientific development, but at the time, they reinforced Darwin's acceptance in biology of the very mechanistic view that he adopted from Lyell. Figure 3.2 gives an idea of how the relationship might have been seen in the 1830s.

	theological	metaphysical	scientific
Geology			Hutton, Lyell, early Darwin ↓
Biology	Paley	Lamarck	Darwin

Fig. 3.2. One view of Auguste Comte's view of the three stages of scientific progress, applied to geology and evolutionary biology. Physiologists were well advanced in giving good experimental evidence for their conclusions such as circulation of the blood or the exchange of gases in photosynthesis and respiration and so could be considered 'scientific'. In contrast, in some other areas of biology such as the origin and relationships between species, biology had to await the long timescale (suggested by geologists) before proposing testable mechanisms.

As outlined previously, Darwin started as a geologist in the tradition of the two Scottish geologists James Hutton and Charles Lyell, who were (in the sense of Comte) in the scientific tradition within geology. Past events had to be explained in terms of mechanisms (or laws) that could still be measured and tested in the present. It is this approach that Darwin would apply to biology and explains his interest in the work of plant and animal breeders and in inheritable variation in populations, as well as his calculations of the potential increase in the number of a species, the large number of experiments he carried out (particularly on plants), and generally trying to make and test predictions. He wanted his mechanism of evolution to be 'scientific'—he considered zoology to be in the 'theological' stage in the mid-1830s. The conclusion is that Darwin developed his method of scientific reasoning as a geologist and then applied it to biology. The fields were not as separate as they are now, and people studying 'natural history' usually had expertise in both areas.

Another influential person for Darwin was the Belgian physicist, statistician, and sociologist Adolphe Quetelet (pronounced 'kettle-ay'

(1796–1874), see Porter 1986). It was from Quetelet's writing that Darwin appears to have realized the potential power of probabilistic mechanisms, especially when both large numbers and long timescales were available. Quetelet had realized that even if some human acts were not 'rational', if there were large-enough numbers of potential events, then good statements could be made about the average number of events, say, per year. Three examples might be the number of murders in Berlin, suicides in Paris, and misaddressed mail in London (the number of letters that ended up in the 'dead letter office' because the letters could not be delivered to the addresses stated). Each action (murder, suicide, or misaddressed mail) could not, according to Quetelet, be considered a rational act—but if there were enough events, then it was possible to make predictions about the expected number of events per year. Quetelet was also one of the founders of the London-based Statistical Association. He perhaps was the start of allowing chance or probabilistic events into science. For the benefit of analysis, we will distinguish between stochastic (chance) and deterministic events; earlier science had really only considered deterministic events.

When Darwin read Malthus's *An Essay on the Principle of Population* in 1838, he accepted the final link in suggesting a mechanism for evolution—natural selection of inherited variability. However, he was certainly primed. Lyell's *Principles of Geology* had quoted De Candolle many times about competition. The potential for a continued increase the in number of individuals in a species (plant and animal) had been recognized by several authors, but it was Malthus in the early nineteenth century who emphasized it for human populations as well. He was a controversial figure in the early nineteenth century because he used his calculations to argue that the human population would inevitably outstrip food resources, and therefore nothing (according to Malthus) could prevent human misery. Darwin writes,

> *In October 1838, that is, fifteen months after I had begun my systematic enquiry, I happened to read for amusement Malthus on Population, and being well prepared to appreciate the struggle*

for existence which everywhere goes on, from long-continued
observation of the habits of animals and plants, it at once struck
me that under these circumstances favorable variations would tend
to be preserved, and unfavorable ones destroyed. The result of this
would be the formation of new species. Here, then. I had at last got
a theory by which to work (1969: 120).

There are two points to be made about this quotation. The first is that he almost has to apologize to his family for even reading Malthus. Malthus was a well-known conservative, and Darwin's family were very progressive (Desmond and Moore 2010), and he wrote his autobiography for his family. But perhaps the main point is that several authors had been aware of, and commented on, the 'struggle for existence'—so he was well aware of its existence, and thus the phrase *'and being well prepared to appreciate the struggle for existence'* is very important. What he appears to take from Malthus is the mathematical precision that guaranteed that with time, the exponential increase in numbers would inevitably lead to competition. Indeed, by seeing the possibility of 'improvement' (through selection), his conclusion was in many ways the opposite of Malthus. Following on from Malthus, many researchers have emphasized the 'selfish' aspect of evolution, but 'cooperation' is also essential for evolution (Penny 2014), particularly for the origin of life. Several authors have pointed out that Darwin was also aware of work by Mathew (1831), who had studied the diversity of plans but apparently not realized the importance of his discovery.

Although Darwin produced summaries of his theory in 1842 and 1844 (the second is well over 100 printed pages), the complete theory was not developed until early to mid-1850s—before then he was apparently concerned that natural selection would not automatically produce the best possible adaptations (see later). Only later did he appear to accept that organisms were less than perfect and that this indeed was a powerful argument in favor of evolution (compared with, say, intelligent design). The theory was not published until 1859 and only after Alfred Russel Wallace had independently come to very similar conclusions about a mechanism for evolution. In the early 1950s, Wallace independently arrived at both

> the theory of descent, and
>
> the theory of natural selection.

Wallace (like Darwin) had extensive experience in natural history outside Europe, though first in the Amazon, then in the East Indies (the Indonesian region). Malthus was again important, but Wallace had read Malthus many years earlier. The theories of the two co-discoverers were similar, though there were some significant differences. Wallace was more of a hard-line selectionist, and he excluded the human brain from a natural explanation. Thus he could not accept that the forces acting in microevolution were sufficient (see later) to account for all of human evolution.

So here we consider the structure of Darwin's theory, divided into twenty points. However, it is also useful to consider the components divided into three subsections, and for the purposes of discussion, I divide them into three main themes. These are (a) the microevolutionary processes, (b) macroevolution (whether evolution has indeed occurred), and (c) whether the processes of microevolution are sufficient to explain all of macroevolution. Distinguished evolutionary biologist Ernst Mayr (1991) used the title *One Long Argument* for one of his books on the history of evolutionary thought. The title comes from a phrase that Charles Darwin used when discussing his own work, *On the Origin of Species*—that his book was one long argument from beginning to end. He saw an overall structure to his book as one long reasoned argument, primarily against Paley's *Natural Theology* (1802). There is a *continuity* of living forms through geological time, and this gives a unity of life in that all living species shared a common ancestor somewhere in the distant past. This arises from the processes of microevolution. The first of the twenty points follows.

1. All populations have the potential for an exponential increase in numbers.

This potential to increase in numbers was well known before Darwin. As we have seen, the second volume of Lyell's *Principles of Geology* (1831) gives many examples of the potential for rapid increase in the number of a species, and he takes his information from earlier authors, including authors from Scotland (Hutton), Switzerland, and southern France (De Candolle). However, by itself (without considering inherited variability), it did not lead to the idea of species changing through time. We have already seen some of Lyell's comments, basically from De Candolle. Like Malthus, these other authors had not seen the possible implications when combined with inheritable variability. The potential for the exponential increase in a population gave an inevitability to this aspect of Darwin's mechanism. This was important in showing that at least part of his theory met the expectations of scientists in 1830s that science should be deterministic; the potential for an exponential increase in numbers meant there was competition for resources, whether we liked it or not (see the next point).

2. Because of limited resources, some competition is inevitable.

For most species, the observed numbers, averaged over a number of years, is approximately constant even though the potential for exponential increase in numbers was always there. Environmental resources are always limited (such as incoming light energy per unit area for plants), and therefore there must be intra- as well as inter-specific competition for resources. This relative constancy of numbers had also been discussed at length in the second volume of Lyell's *Principles of Geology*. Taken just by itself, it had not led to the idea of any change through natural selection. Lyell's main discussion was in terms of competition between different species. As we have seen, he had quoted De Candolle's *'all plants are at war, one with another'*. But Lyell had certainly also recognized competition within a species.

3. Some of the variability within populations is inherited.

New variability is continuously arising, regardless of any 'needs' of the organisms. At the time of Charles Darwin, biologists did not know the details of genetics or of DNA, though they did know generally about inheritance. Perhaps the best explanation of why other biologists did not recognize Gregor Mendel's advance that turned 'inheritance' into 'genetics' was that they had thought that inheritance should also explain development (Sandler and Sandler 1985). And as we have seen, another inhibiting factor for some, but not all, biologists was the acceptance of the idea of an 'unchangeable essence' for a species. Any observed inherited differences, under that hypothesis, were that any deviations from the ideal type would eventually be lost.

Box 3.1
A - NATURAL SELECTION - (MICROEVOLUTION).

Increase in population numbers (an ecological component).
1. All populations have the potential for an exponential increase in numbers.
2. However for most species the observed numbers averaged over a number of years are approximately constant. Thus, because environmental resources are limited there is intra- as well as inter-specific competition.

II. Inherited variability (a genetic component).
3. There is variability within populations, some of which is inherited. New variability is continuously arising, regardless of any 'needs' of the organisms.
4. Some of this inheritable variability can affect the probability of survival and reproduction in the present environment (physical and biological). The probability of surviving and reproducing can increase or decrease relative to other members of the population.

III. Natural Selection
5. Processes I and II can lead to a change between generations in the frequency of different genetic variants, this is microevolution (and the probabilistic component is emphasized here). Thus there is continuity between generations.

That individuals varied from both environmental and inherited differences was known from the seventeenth century, but the variation was often considered unimportant and even undesirable—it was deviation from the 'perfect form' (essence) of the species. The job of the taxonomist was (on this view) to discover the original perfect form being masked by variation.

Thus the task was to 'ignore' the variation; variation was an indication of the imperfection of the world (why an all-powerful Creator would make such an imperfect world was a difficult question). Okay, I guess they wanted the system to evolve. Lewontin (1974) says the following:

> *The metaphysical introduction of ideal bodies moving in ideal paths, so essential to the proper development of physics and so consonant with the habits of thought of the 17th century, was precisely what had to be destroyed in the creation of evolutionary biology. Darwin rejected the metaphysical object and replaced it with a material one. He called attention to the actual variation among actual organisms as the most essential and illuminating fact of nature. Rather than regarding the variation among members of the same species as an annoying distraction, as a shimmering of the air that distorts our view of the essential object, he made that variation the corner-stone of his theory. Let us remember that the Origin of Species begins with a discussion of variation under domestication (p. 5).*

4. Some of this inherited variability can affect survival and reproduction.

This refers to survival and reproduction in the present physical and biological environment. The probability of surviving and reproducing can increase or decrease relative to other members of the population. In one sense this appears as quite a weak statement—'some' inherited variants and the 'probability' of survival and reproduction. But there are large numbers and plenty of time, so some change is inevitable. Given the earlier view that variability was undesirable, many authors had recognized that selection would tend to eliminate these forms; in current terminology this is negative selection. However, the realization that plant and animal breeders were able to use the variability to give new breeds of domesticated species allowed a new view of this variability: domesticated breeds were 'improvements' (from the human viewpoint), so variation was in some sense 'natural' and offered at least the potential for increased survival of some members of the population. Perhaps a qualification could have been made that many domestic varieties tended to revert to a more natural breed if the plants or animals escaped from domestication. This did not counter the argument that change was possible (as opposed to many essentialist views).

Darwin also pointed out that morphological changes that occurred (for example, in domestic dogs or in the cabbage/cauliflower/broccoli groups) were far greater than what normally occurred between *genera* even though they (dogs or broccoli and cabbage, as judged by interbreeding) were still in the same species. Consequently, the morphological changes selected by plant and animal breeders were more extensive than what occurred between species in nature. As we would say now, variability is a resource. It was clearly an important step to give up the theory of an ideal form of a species, but it is not clear when Darwin did this. Possibly the experience from plant and animal breeders was sufficient?

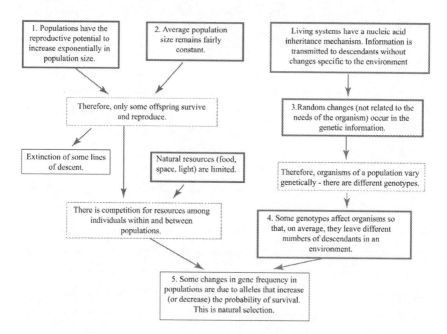

Fig. 3.3. Another way of looking at microevolutionary processes. The left-hand side and center are more ecological, and the right-hand side more genetic.

5. Some change in gene frequencies between generations is inevitable—natural selection.

Processes 1–4 (microevolutionary processes) lead to a change between generations in the frequency of different genetic variants: microevolution. The probabilistic component is emphasized here and was a novel part of his proposed mechanism. As mentioned, there is a large number of individuals and long periods of time, and so some change is inevitable. In addition, it is now emphasized that there is continuity of gene frequencies between generations. We can't stop evolution—dammit! Those little influenza viruses (RNA viruses, see Eigen limit later) just keep evolving year after year. Every northern winter, there is a new selection of the viruses used for antibody production; and every southern winter, there is a new selection again. We really can't stop these little beasties evolving!

Given the background knowledge (components 1–4), it is an almost inevitable outcome that there will, over time, be changes in the population—and that the population could in some sense be better suited to its immediate environment. This process would result just from (in Lyell's words) 'causes now in operation', processes which could be studied in the present by experiment and observation. This was a major triumph of Darwin's theory—it allowed continuing adjustment to the environment without postulating mysterious, unknown, and unobservable mechanisms (either operating currently or at some remote time in the past).

There needs to be a word of caution here: the mechanisms for generation of new inheritable variability were unknown in Darwin's time, and this problem was not solved until well after Mendel's laws and then DNA were discovered, and we still do not know how a particular mutation might cause a specific change in selection. Nevertheless, it was known in the nineteenth century that variability did occur and that it increased after sexual reproduction, but less after asexual reproduction. In addition, some of the variability was 'useful'—at least for artificial selection (new varieties). Consequently, his theory depended only on observable features, including new inheritable variability.

To sum up this section, it is important to note that 'natural selection' is the consequence of several processes (1–5); and when there is any ambiguity, it is clearer to speak of '*the processes of natural selection*' rather than just 'natural selection'. In many ways, it is more accurate to say, for example, 'the processes of natural selection leads to a better fit between organism and environment' rather than 'natural selection leads to a better fit between organism and environment'.

As a shorthand, the latter is fine—as long as it is clear it is just that. It is misleading to treat natural selection as if it were almost a thing or an object and not a process.

Box 3.2

B - MACROEVOLUTON
6. Species can split into two or more lineages
7. Extinction of species has occured
8. A very long time scale is available for evolution
9. The principle of divergence - descent with modification

---

6. Species can split into two or more lineages.

As we discuss in the first chapter, traditionally (certainly before1600) it was assumed that life continued to arise spontaneously and that some forms of living matter (and non-living) could interchange (werewolves, for example). But by the early eighteenth century, the idea of species having permanence through time became well established. But as we have seen, it was often combined with the ancient Greek idea of 'unchangeable essences'. Accepting that there were no unchangeable 'essences' or 'forms' allowed evolution to occur and also allowed a potential increase in the number of species.

Only one point needs to be made here. It was not obvious to all earlier biologists that a species could split into two or more lineages. An important inhibitory factor was, as we have mentioned, the ancient Greek concept of the unchanging 'essence', 'type', or 'ideal form' for any class of objects, including species. The essence was, by definition, unchangeable. These Greek ideas had been integrated into medieval 'Christian' theology, and it was now assumed (by definition) that the essence or type could *not* change, let alone split into two. Strictly speaking, the early Greeks had a variety of views; it was the ideas of Plato and Aristotle which were adopted by the early-medieval theologians. Some Greeks had other views. *'I can see horses, Plato,'* said Antisthenes. *'But I nowhere see your ideal horse'* was the retort of another Greek scientist/philosopher who thought it unnecessary to postulate un-seeable ideal forms—we could 'see' horses and measure them, etc., but we cannot see (or measure) the 'essence' of a horse. Go, Antisthenes!

7. Extinction of species has occurred.

Given the present-day recognition that there have been massive and ongoing extinction of megafauna (and many other species) over the last 40,000–50,000 years, it may seem surprising that in the eighteenth and early nineteenth centuries, it was uncertain whether extinctions occurred at all. In earlier times, it had been assumed that all forms of life continued to form spontaneously 'from the earth'. Thus there was really no question of extinction being possible; organisms could regenerate. If all the living forms of a 'type' died, it would still be expected (on that view) that new forms would arise spontaneously. It was 'known' that the earth or soil in England was different from that in Ireland—grass snakes developed spontaneously from the soil in England, but not in Ireland. Case proved!

An argument against extinctions arose later in the eighteenth century after the concept of biological species had developed. It went something like 'species are created by an all-powerful Creator to fit their environment, and surely no perfect work of the Creator will be destroyed, particularly as they fit their environment perfectly'. Even though remains of unknown species had been found only as fossils, it was still possible that living examples would eventually be discovered in unexplored parts of the earth (especially in the deep ocean if the fossils were marine). Others argued that it was no business of mere mortals to put limits on the Creator, who could do as s/he damn well pleased.

By the early nineteenth century, some naturalists had accepted the argument of a series of creations, accepting the Creator's divine right to destroy living things and create new ones; these scientists accepted extinction. Scientists such as Cuvier and Lyell, though very different in their views, were initially non-evolutionary but accepted extinction— Cuvier through catastrophes and Lyell through gradual extinctions according to ecological laws. Others such as Lamarck (and probably Geoffroy Saint-Hilaire) were pre-Darwinian evolutionists but denied extinction occurred. To Lamarck, it appears species automatically had the ability to adapt to new conditions. If we couldn't find fossil taxa living in the present, it was because they had adapted and changed to new conditions, and we didn't recognize them yet. It was difficult for these early evolutionists to accept extinction because their 'mechanism' of evolution was just the (metaphysical) statement that species could automatically change and 'improve'. So even if some scientists were 'correct' about species evolving, they were simultaneously 'wrong' about how it occurred. Pattern and process are both important.

From our viewpoint today, the non-evolutionists Cuvier and Lyell had high scientific standards but were willing to put limits on science, outside of which the Creator had a role; Lyell did later accept that evolution had occurred. On the other hand, the evolutionists in the late eighteenth and early nineteenth centuries were 'unscientific' by our standards because they made no attempt to test their speculations on mechanisms by which evolution may have occurred. In modern terms, a theory that claims to explain every conceivable event in fact explains nothing (it can't be falsified or disproved, a key Popperian requirement for a scientific theory). Darwin came from within the tradition of Lyell and was able to maintain the high scientific standards of only proposing mechanisms that could be studied and tested in the present.

8. A very long timescale is available for evolution.

We have already outlined this question in the previous chapter. The evidence against long-term evolution did look quite strong when French scientists (following Napoleon's conquests) studied mummies of animals in Egypt that were about 4,000 years old. The mummies of cats, dogs, crocodiles, or ibises, for example, contained animals that looked just like modern species! If the world was only around 6,000 years old and the species 4,000 years ago were already fully modern, then it was stretching incredulity to suggest that all of evolution had occurred in a mere 2,000 years (and by processes that could scarcely be imagined).

The age of the earth was still controversial in the early nineteenth century. Nevertheless, in the early editions, Lyell did not specify how old he thought the earth would be. Darwin thought that the mechanisms of evolutionary change (natural selection) would require a long timescale of hundreds of millions of years. A scientific attack on this very long timescale started from the 1860s (after the publication of *On the Origin of Species*) and was from the physicists (particularly Lord Kelvin) who argued for a shorter timescale (though still tens of millions of years) because of the argument that can be paraphrased as 'the sun couldn't be that old, or it would have run out of oxygen to burn for energy'! The discovery of radioactivity at the end of the nineteenth century allowed the sun a much larger and longer-lasting source of energy and eventually allowed a radioactive dating of the age of the earliest rocks. Early in the twentieth century, the New Zealander Lord Rutherford was among the first to suggest that natural radioactivity could be the source of energy for the sun and that this would allow for an older age for the earth. The age of the earth is now put at about 4,600 million years ago with the origin of recognizable forms of life possibly occurring 600–700 million

years later. But we still do not know if they were before or after the Last
Universal Common Ancestor (LUCA).

9. The principle of divergence—descent with modification.

The splitting of lineages has continued over geological time. Thus
there is a continuity of living forms through geological time, which
gives a unity of life in that all living species shared a common ancestor
in the past. This arises from the processes of microevolution that were
discussed earlier. Today we are used to seeing relationships between
taxa as a tree (or network) that we forget that from the time of the
ancient Greeks, it took over 2,000 years of studying plants and animals
before a tree diagram was proposed in its present form. An evolving
tree is not a self-evident pattern for relationships among plants and
animals. We saw earlier that the real problem of the relationship of
species only became important once species were assumed to have a
continuous existence over long periods of time. This is after the idea
of continuous spontaneous generation became improbable, and it was
no longer accepted that one form of life could 'transmute' into a quite
different form during the lifetime of an individual (no werewolves!).

It was certainly not apparent to scientists in the eighteenth and early
nineteenth centuries that a rooted tree diagram (Figure 3.1) was a good
way of indicating relationships between species. In all the sciences, it
required major efforts before fundamental patterns that allowed the
development of mechanistic hypotheses were recognized. For example,
in astronomy, there were ellipses describing the movement of planets
around the sun. That the planets (the wanderers) behaved differently
from the fixed stars was certainly recognized by 3000 BC, but it took
around 4,500 years before Copernicus found the pattern to describe
their motion—rotating in elliptical orbits around the sun in the same
direction and the same plane (with the moon going around the earth).
In chemistry, it was Mendeleev's table that showed the relationship of
the chemical properties of the elements; in one aspect of geology, the

importance of matching the coastlines led to the ideas of continental drift and plate tectonics. In all these cases, as well as in using trees for relating biological species, the answer seems 'obvious' to us today. In each science, it was a major achievement to finally recognize the patterns in the data that led to the most powerful hypotheses about the mechanisms (processes) involved.

There are many times where a tree will not be a good model for describing relationships, and probably a network pattern (that allows cycles in the graph) is required. The most obvious cases are plants where hybrids are common. Another well-studied case is the endosymbiotic origin of both chloroplasts and mitochondria. Genes in higher plants come from at least three different sources (admittedly a long way back): nucleus, mitochondria, and plastids (chloroplasts). For this question, a network is an appropriate model, though in practice each gene tends to be considered separately and a tree drawn from each. However, given the random nature of mutations and the random nature of selection, we would be very surprised if every tree was identical!

Within eubacteria, the transfer of some genes between bacterial 'species' occurs through 'lateral transfer' by plasmids and other mechanisms. Lateral transfer can happen within eukaryotes—for example, between an insect and its host via parasites. A different case is when there is adaptation of a small population (from within a larger species) to a new environment, and it is difficult to represent this by standard methods. Examples include polar bears possibly arising within brown bears and some high alpine buttercups in New Zealand, which arise from within widely dispersed species from lower altitudes. Going back much earlier to the origin of life, the best models we have at present assume a large number of small genetic fragments that would continually recombine (see later 'The Origin of Life'). None of these cases overrides the use of the tree relationship between eukaryotic species as a good useful model. However, the model of interbreeding species may not be so useful outside the sexually reproducing eukaryotic groups, although in this case lateral transfer could be considered equivalent.

***Components 10–20. The sufficiency of microevolutionary processes—
Darwinism or actualism.*** The processes of microevolution are sufficient
to account for macroevolution (though remember point 20 at the end).

10. The processes of microevolution are sufficient for macroevolution.

Within science, the first two major sections of macro- and microevolution
are now fully accepted; evolution has occurred, and the processes of
microevolution are necessary for evolution. It is decreasingly controversial
that the mechanisms we are able to study in the present are sufficient to
account for past changes. As stated above, this question is very different
from whether we know all the mechanisms—we do not (see point 20
later). Rather, it is whether studying 'causes now in operation' (including
extraterrestrial impacts) will be sufficient to explain past changes or whether
we appeal to some type of mechanism that cannot now be studied—and
to that extent would be still outside the realm of normal science.

What is important is the continuity between generations—'numerous
slight successive changes'. Darwin made a bold statement:

> *If it could be demonstrated that any complex organ existed, which
> could not possibly have been formed by numerous, successive, slight
> modifications, my theory would absolutely break down. But I can
> find out no such case* (1969: 189).

Examine this statement carefully; it is the continuity, not equality of
rates, which he saw as critical to his theory. This statement is still essential
to current work on the sufficiency of microevolutionary processes.

One of the most interesting questions here is how to get from the
DNA genotype to the phenotype as actually observed. The discussion
usually centers on the difference between a 'blueprint' and a 'recipe'. In
an architectural blueprint, for example, there is more or less a one-to-
one relationship between the blueprint and the actual building. Given

a building, you could probably reconstruct a good blueprint; given a blueprint, you could construct the building. However, we cannot (yet) do this with the DNA sequence of a genome, and a recipe is currently a better analogy. A recipe should be sufficient information to make the end product. However, it is not currently possible to infer either what slight differences to the recipe (DNA sequence for the organism) would make nor to infer the recipe precisely given the finally cooked product. Okay, you might be close each time, but there is not (yet) a simple one-to-one relationship. However, in the present context, the sequence, given it is transcribed, processed and translated in a cell with its own information is sufficient for the organism to develop. Evolution has occurred; mutations, population increase, and selection can be shown in the laboratory and in the field.

11. Microevolutionary mechanisms are sufficient for the origin of humans.

In particular, microevolutionary mechanisms are sufficient to account for the origin of humans, including their mental powers, language ability, and social, religious, and ethical systems. Right from the beginning of his study on the origins of species (in the late 1830s), Darwin included humans in his theory. He made extensive notes on humans, including the development of physical and mental abilities of young children (his own), although it was not until the 1870s that he published the results of this work. So in *On the Origin of Species*, the only reference to humans was right at the end when he said simply, *'Much light will be thrown on the origin of man and his history'*. Later he wrote the book *The Descent of Man*, which was devoted entirely to humans, and concluded, *'Man, like every other species, was descended from some pre-existing form'*. His book included almost everything about humans—morphology, emotions, speech, and reasoning. However, by this time, other authors such as Lyell had already written books on the antiquity of humans. Lyell's book on the antiquity of humans was part of his working through Darwin's reasoning and coming to accept the main evolutionary conclusions.

With respect to humans, a dominant view in the nineteenth century had been that of the early mid-seventeenth century French scientist and philosopher René Descartes (whom we mentioned earlier), whose theory was of a gulf between humans and all animals. As mentioned in the first chapter, Descartes concluded there was 'material' matter and 'mind' matter. According to this Descartian view, the human mind was quite different from that of any animal. They (and basically everything except the human mind) should be considered as machines or automata, responding to stimuli in the environment. On this view, the realm of science was the material world (which theologians should leave to scientists); conversely, the realm of human mind and speech was not the realm of science and should be left to theologians!

This 'dualist' view is certainly non-evolutionary in a Darwinian sense because it hypothesizes an unbridgeable gap between humans and all other forms of life. It has already pointed out that earlier (eighteenth century in particular), the Great Chain of Being had been a dominant view, and then it was assumed that there should be a continuous series of animals up to and including humans (and this would have included mental attributes). Indeed, when the great apes (chimpanzees, gorillas, and orangutans) were first discovered (in the seventeenth century), their close similarity to humans was considered to support the Great Chain of Being (this is another form of the principle of continuity).

There were probably major political and religious advantages in accepting Descartes's view of humans; it established science as an independent discipline not subject to the theologians (who were a pretty brutal lot who burned people at the stake if they disagreed with them; see Pinker 2011 for the reduction in violence over time). From the eighteenth century onwards, Descartes's 'scientific' distinction of the human mind was accepted almost as axiomatic, so much so that when Darwin was a student at Edinburgh, a student scientific society struck out from its minutes all reference to talk by a member who advocated a 'materialist' view—that is, the mind is dependent on material (physical) matter. This attitude to the human mind was probably the only major difference between Charles Darwin and A. R. Wallace because Wallace eventually became a spiritualist and accepted that 'something else' was required to explain the origin of the human mind.

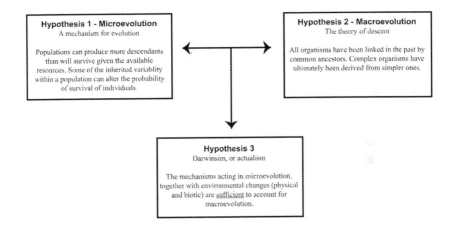

Figure 3.4. The three main components of Darwin's theory. They are shown with arrows in each direction because they are mutually supporting in that, for example, the theory of descent is supported by the existence of a mechanism that could lead to species modification and divergence; but the theory of descent has also lead to the search for mechanisms that would result in descent with modification.

Darwin accepted the work from comparative anatomists (such as Thomas Huxley (1825–1895)) that chimps and gorillas were the most similar primates to humans. Darwin's conclusion was that many animals showed, in at least a minor way, features more strongly developed in humans. Some of his work helped lead to psychological and behavioral studies on animals, though it was not until 100 years later (in the second half of the twentieth century) that the subject really took off. It has only been in the last forty years that the close phylogenetic relationship between great apes and humans was established and even more recently their mental abilities studied scientifically. The evolution of humans is the subject of Chapter 5.

12. The principle of continuity: nature does not make jumps,

Basically, there is a continuous series of intermediates between generations. This is often called 'gradualism' (a much misunderstood and misleading term; see the previous chapter). Here we emphasize just the continuity and overlap of forms between generations, each change between generations being reversible in principle. Certainly, the rate of change between generations may be variable. Changes in gene frequency occur on a biological/ ecological timescale rather than during the lifetime of an individual (although mutations occur all the time). The principle of continuity and reversibility between successive states is fundamental in many aspects of science. Darwin frequently used the phrase 'nature does not make jumps' ((*natura non facit saltum*) there was continuity between successive states during differentiation, for example). This linked his ideas into other aspects of mathematics and science (including physics). In mathematics, Leibniz had used it in the development of calculus.

The principle of continuity is basic to scientific thought; we can't explain things scientifically by miracles. We can't say 'Suddenly, the ancestral bird had feathers' or 'the long-extinct *Eohippus* will eventually evolve into a horse' and consider it as a scientific explanation. Rather, in sciences such as physical chemistry, many thermodynamics analyses assume continuity between successive microstates. Change between each successive microstate is changing according to known mechanisms. Thus the concepts of continuity and microscopic reversibility are similar; in science, there are no miracles. Reversibility in physical chemistry just requires that a process can be broken down to a large number of microscopic steps, each of which is reversible in principle. In biology, we emphasize the continuity by the overlap of gene frequencies between generations; even though the rate of change between generations may be variable. The gene frequencies in a population at any point in time can be considered the equivalent of the state during a chemical reaction.

Thus it is useful to consider analogies from the physical sciences under continuity (and microscopic reversibility)—that is, a mechanistic step-by-step approach. Evolution is using a standard scientific approach; this was a real contribution from Darwin. Later we will apply the same approach to the origin of life—in the early stages, each step will be a normal chemical one; later it will probably be a genetic or epigenetic mutation.

13. Populations and selection occur at the level of individuals.

Because of the nature of inheritance, the benefit (or loss) went to the possessors of a gene or gene combination (including cooperativity). Selection thus acts on individuals, and populations change in gene frequencies. Any changes are not in principle 'for the good of the species', nor can the processes of natural selection be oriented to 'future goals' (though certainly a given mutation might allow other ones in the future). The processes can lead to cooperation both within and between species, and early biologists were well aware of this possibility. Co-evolution between species occurs regularly in nature, and for a cell to work, dozens of enzymes have to cooperate—and work together.

Individual selection arises from the nature of inheritance and from competition, both inter- and intra-specific. The exact, or even approximate, mode of inheritance was a mystery in the mid-nineteenth century (except perhaps to Gregor Mendel, who published a few years later in the 1860s). As mentioned above, all that was needed was knowledge that there was transmission from parents to offspring; this was the 'known mechanism'—it could be observed repeatedly as occurring even if they did not understand the biochemistry, though the biochemistry helps enormously. The effect of selection was perhaps a major reason leading to the loss of essentialist thinking about species (that required species being held together in some unknown manner). It is hard for us to realize today just how dominant essentialism had been in so many aspects of medieval thinking.

This reliance of the mechanism on 'individual' selection did not in any way preclude cooperation within and between species (Penny 2014). The concept of what we would now call 'co-evolution' had been recognized earlier; and Lyell, for example, had discussed beneficial interactions between species. However, 'political philosophers' and economists often had no knowledge of positive interactions within biology and earlier had concentrated more on the 'selfish' aspects of selection—shame on them.

14. There is a strong chance (stochastic) factor in evolution.

This arises in many ways, including in mutations (origin of inherited variability) and in selection (or lack of it). There are both stochastic and deterministic factors in evolution. Allowing a major non-deterministic role in evolution was an aspect that was seriously criticized in Darwin's time, especially by physicists (who then in the twentieth century 'invented' probabilistic reasoning 'all by themselves'—a fantastic achievement!).

Allowing chance events to play a major role was perhaps the most original aspect of his theory and perhaps the most criticized—'*the law of the higgledy-piggledy*' was the view of one physicist-philosopher, John Herschel (1792–1871). It has already been mentioned that Darwin's realization that strong conclusions can be drawn about chance events if the number of observations is large and came from the Belgian Adolphe Quetelet. As we have discussed, he showed that when there were many events, even the number of 'non-rational' events per year could be predicted.

Early nineteenth-century (and since Newton at least) science was very deterministic. Newton's laws were considered the model of good science and allowed the calculation of the positions of the planets many years into the future or back into the past, provided the initial measurements were sufficiently accurate. In addition, much of theology was also deterministic; extreme forms were 'predestination', where the life of an individual was predetermined or 'preordained'. There was then a general acceptance that 'at the appointed time', an individual was 'called' by their Maker.

The handling of 'chance' within Darwin's writing was sometimes ambivalent as there was always the possibility that what appeared to be 'chance' may just be due to as yet unknown causes. Nevertheless, the recognition that chance occurred in

> the production of variation,
> whether it was advantageous (either now or in the future),
> whether advantageous variation was present when 'needed' (say as the climate changed), and
> even if advantageous, whether it would help a particular individual.

All were features new to science. We will see later that the modern view (neutralism) places even more emphasis on stochastic processes.

An important point to emphasize is that possessing an 'advantageous' mutation is no guarantee it will help in an individual case. This is because there are many other factors affecting survival. Consider an example of two thistles growing on the bank of a stream. One gets swept away in a flood and leaves no descendants. We certainly cannot conclude that the remaining plant was 'better' than the other—only luckier! A selective advantage is averaged over many individuals.

A view of the importance of the introduction of probabilistic reasoning into evolution is given in François Jacob's book on the history of heredity, *The Logic of Life*:

> *Until the middle of the nineteenth century, the living world was considered as a system controlled from the outside. Whether they had been fixed since creation or whether they had progressed through successive events, organized beings were always arranged*

in a continuous series of forms. Apparent breaks in the hierarchy were due to omission, ignorance or inadequate listing. The existing structure of the living world, then, expressed a transcendental necessity. That living beings could be any different from what they are, that other forms might inhabit the earth, this was just inconceivable. The theory of evolution swept away the idea of a preconceived harmony that imposed a system of relationships on organized beings. The necessity for the living world to conform to its present pattern was replaced by the contingency that already governed the sky and inanimate things. Not only might the living world have been totally different; it might equally well never have existed at all. Organisms became components of a vast system of higher order embracing the earth and everything on it. The form, properties and characters of living beings, therefore, were subject to regulation from within the system (1973: 172–173).

We will come to François Jacob and his discoveries of how DNA is controlled in the next chapter; here it is sufficient just to give his views on what happened during the nineteenth century. But the main point here was that the probabilistic reasoning of Darwin was a novelty at the time.

15. Adaptations and new features explains the fit between the organism and its environment.

Highly ordered structures can arise gradually from less ordered states and lead to the origin of new features: orchids can eventually evolve from chlorella-like algae, mammals and birds from unicells. Each intermediate form must not be harmful to its possessor. It all happens over hundreds of millions of years!

The mechanism explains the overall fit between form, function, and the environment. Any slightly improved variant (that is inherited) that tends to fit the environment better is more likely to survive. This point was critical to the theory; it explained the apparent design in nature. This is in many ways the key point in understanding the consequences of evolutionary theory. Darwin showed that it was possible, in principle, for natural mechanism to explain the overall 'fit' between function, form, and environment. Any slightly improved (inherited) variant that tended to fit the environment better was more likely to survive. Given time and repeating the processes over and over again provided an alternative explanation of the 'apparent' design of organisms.

This process of 'adaptation' has resulted in a lot of arguing over wording. Sometimes people will say something like 'Long teeth are an adaptation for carnivores'. Using '*for*' in that expression raises difficulties. It sounds almost as if there is 'purpose' in having the long teeth (or whatever other feature is being considered). Mouth parts of insects are adaptations 'for' feeding; sunken stomata are adaptations 'for' reduced water loss in desert plants. This usage is really only a problem over the use of words; we are using a shorthand expression for a much longer statement. We could say something like 'Over millions of years, there has been an increased tendency for carnivores to survive if they have slightly longer teeth', etc. Although 'correct', it is rather tedious—like being politically correct all the time. You do have to be careful in evolutionary studies not to get distracted over problems of wording. Remember the story about the famous evolutionist John Maynard Smith. On arriving at morning tea one day, he found an intense discussion going on. *'Are you arguing about the world?'* he asked. *'If so, I am very interested. Or are you arguing about the use of words? In which case, I am not interested at all.'* It is necessary that a word in any language has several shades of; meaning, we have to be clear which usage we are making and then get on understanding the world and how it works.

There used to be considerable argument over whether the mechanism of numerous slight modifications being selected over long periods of time would really work. Nowadays the process can be carried out in a computer or in a test tube or in molecular medicine. Computer scientists find that using 'evolutionary programs' or 'genetic algorithms' are very good at solving quite difficult problems. They can follow a biological example and have a series of alternative versions of the program, each with a set of required parameters (in other words, a population with variation). Then the program evaluates how each set of parameters works in a particular case. This is followed by a round of replication where successful versions of the program have an increased chance of being duplicated, and less useful versions have lower chances of being duplicated. Then there can be some 'mutations' where a parameter can have its value modified. In some cases, there is recombination between sets of parameters—new sets are made by combining parameters from two groups. Then the whole process is repeated over and over and over again. It works! Just like biological evolution, the programs can get better but are not guaranteed to get the absolutely best result—just like in biology.

Similarly, there is an important new field of in vitro evolution where RNA molecules evolve in a test tube. The starting mixture is of a very large number of variants of RNA molecules (made synthetically with at least some positions randomly adding any of the four nucleotides). This population of RNA molecules is tested, for example, by passing through a column that has a compound you require the RNA to bind to. The RNA molecules that bind slightly to this compound are retarded by the column. These RNA molecules are then collected and then replicated several times by a process that allows new mutants to arise. The new population of RNA molecules is tested again and the process repeated. All sorts of new reactions catalyzed by RNA are found—'evolution in a test tube'.

There are many similar cases of short-term evolution. One example is that in some cancers, cells increase their resistance to chemotherapy over time. In some cases, there may be a duplication of a gene that gives a slight resistance against the drug being used for chemotherapy, and so these cells grow slightly faster than the others. Having two copies of the gene slightly increases the chance of another duplication of that gene, which gives somewhat more resistance, which means the cell grows slightly faster and so on. Another well-known case is the 'superbugs' that are bacteria resistant to a large number of antibiotics.

In other cases, bacteria or yeast are grown on a new medium that has a compound (source of energy) that the organism is not able to use. As long as a low level of normal sugar is supplied (which keeps the organism alive and growing), mutants often arise in some cells, which allow the new compound to be used slightly. This can continue, with new mutations allowing even better growth. Experimental evolution with microorganisms is a very active research field. It would take much longer to do similar experiments for organisms with a long life cycle, elephants or many trees, but the principles are now established. The mechanism proposed by Darwin works in practice.

An example of the type of problem pointed out by critics is given by the evolution of feathers. It is difficult to imagine a series of intermediates between feathers and reptilian scales where the intermediates would assist flying. Surely, the argument went that feathers would be no use at all for flying until they were fully formed. And if the intermediates were of no use, then they could not be acted on by the processes of natural selection. (Unless there was an advantage to the organism, there is no selection.) Some people then suggested 'macromutations', where feathers appeared, fully formed, all at once. This has been called the 'hopeful monster' theory, and it is not referring to 'a cause now in operation'—we did not see such mutations today, and even if we did, the 'bird' would not have all the other prerequisites (wings, musculature, etc.) needed for flying.

The usual explanation is that feathers did not arise 'for flying', but for insulation (and that they were a useful 'pre-adaptation' for flying). If early feathers increased insulation (and thereby aided temperature regulation), then intermediate states between scales and feathers would benefit their possessors. The best evidence comes from fossils of early 'feathered dinosaurs'—small theropod dinosaurs that are the ancestors of birds. We now say that feathers were 'co-opted' or 'recruited' for flying from their original function in insulation. And with gene duplication, one copy can retain its original reaction, and the other may get recruited into a new function.

16. A lack of perfect design.

Many features are expected to be less than fully perfect for several reasons; constraints from pre-existing forms, limited genetic variability, conflicting requirements, and all intermediate forms must be functional. Natural selection is expected to often lead to local optima, but not necessarily to the global optimum. Examples considered are the explosive increase in the number of some plants and animals introduced into new lands (the previous inhabitants did not appear to be perfectly adapted to their own environment, though there may be other reasons, such as new species escaping from their natural parasites). Similarly, eyes (including humans) did not give perfect vision, and males had nipples on their breasts (which was unnecessary and therefore imperfect to the Victorian mind). Adaptation is relative rather than absolute.

Darwin had the outline of his theory in 1838 but did not publish the theory until 1859. One suggestion (Ospovat 1981) for the twenty-year delay was that he did not initially see how variation that would always lead to a perfect fit between the organism and its environment would arise. There did not seem to be a mechanism that would lead to the right variation at the right time. It was only in the 1850s (according to Ospovat) that Darwin realized that

1. variability being generated regardless of the 'needs' of the organism could not guarantee that useful variability would be available when 'needed', but
2. organisms (including humans) were not perfectly adapted to their environments.

In other words, organisms were not perfect, nor was there a mechanism to guarantee perfection.

Previously it was almost axiomatic that organisms were perfect in relation to their environment; surely no all-powerful Creator would make anything less than perfect! This lack of perfection then became one of the most powerful tests between his evolutionary model and the earlier creationist biological model. Many examples could be given, but the lack of absolutely perfect vision in humans is one (birds have better vision, so it is possible). There are innumerable examples of lack of perfection, including the following:

> development of cancer in animals
>
> calcium loss from bones, particularly in older people
>
> rubisco and oxygenase, affinity of the enzyme for both CO_2 and O_2 limits the rate of photosynthesis
>
> lack of cellulase in most animals
>
> lack of nitrogenase in eukaryotes
>
> astigmatism and sight, many animals have better vision than humans
>
> need for vitamins, including essential amino acids in humans and vitamin C in humans and guinea pigs (intelligent design would surely produce all these compounds within the organism)

Indeed, the lack of perfection is the rationale of genetic engineering— for example, resistance of plants to insect herbivores. It is now almost axiomatic that we can 'improve' on some aspects of nature; this attitude was certainly not true in the early eighteenth century, when it appeared axiomatic that organisms fitted their environment perfectly.

17. No predetermined pathways of development.

As mentioned earlier, the two main approaches before Lyell and Darwin were catastrophism and orthogenesis. Orthogenesis in particular suggests that organisms, over long periods of time, slowly unfold over predetermined pathways. If, for example, all vertebrates were removed from the earth, then under orthogenesis, they could develop again. In contrast, under actualism (Darwinism), there is no expectation that would evolve again; it all depends upon chance mutations. This conclusion (no preformed pathways of development) reinforces the earlier one about no ultimate causes. Simple physical principles will limit the types of organisms that evolve, but there are many body plans that fit physical necessities.

A common phrase to describe the issues is 'contingency'. The actual organisms we observe are 'contingent' on many events in the past, nor can we predict the order of events into the future. We can often predict general events, such as a mutation that might give resistance to herbicides, even if we cannot predict which mutation will arise first. The overall conclusion helps distinguish Darwinism from orthogenesis that assumed predetermined pathways of development.

18. It is not necessary to postulate a plan, or ultimate cause, for either organisms or the universe.

There need be no 'ultimate purpose' to life, including that the world was made for humans 'to have dominion over'. The concept of ultimate cause had been very influential in Western thought (and perhaps in most cultures), at least from the time of ancient Greek philosophers and probably much earlier. Darwin's theory does not exclude the possibility of purpose in the universe, only that it is unnecessary to postulate it.

The lack of the need to postulate a purpose behind biological (and physical) processes was perhaps a major implication of evolutionary theory and possibly why evolution had so much impact in both science and the general community. Previously, one of the strongest themes in biology had been teleology—organisms were designed 'for a purpose'. This united science and theology and has already been referred to under natural theology. If the origin of complex organisms, and the adaptation of organisms and their environment, could be explained scientifically (by 'known' processes still occurring today), then it was not necessary to postulate a purpose (or teleology). It is now standard biology to exclude teleology, and we tend to forget that this was a major achievement of nineteenth-century biology. The extreme form was to assume that the world was created 'for' humans and animals and plants created for human benefit. A classic example is

> *and God said unto them, Be fruitful and multiply, and replenish the earth, and subdue it: and have dominion over the fish of the sea, and over the fowl of the air, and over every living thing that moveth upon the earth. And God said, Behold I have given you every herb bearing seed, which is upon the face of the earth, and every tree, in which is the fruit of the tree yielding seed; and you it shall be for meat. And to every beast of the earth . . .* (Genesis 1: 28–30).

'Subdue' it, 'have dominion over', and 'I have given you every . . .' are very strong statements to the effect that the world was created for a purpose—for the benefit of humans. These attitudes were challenged by Darwin's theory and could no longer be supported scientifically. The modern conservation ethic is based on this major change of attitude that humans are subject to the laws of nature, not that nature existed for the benefit of humans. A useful fallback position for theistic thinkers is to modify the above extract to have 'responsibility' and 'care' for nature—a big improvement over the traditional view. However, the traditional view was that nature was created for the benefit of humans; it was all part of the purpose of the universe. Modern science has changed this view.

In contrast, the earlier theory of orthogenesis assumed organisms slowly change through predetermined pathways over geological periods of time. However (and as we have said), under Darwinian evolution, if all vertebrates were removed from the earth, there is no guarantee that the same body plans would develop again. Maybe, maybe not. This conclusion reinforces the previous one. Simple physical principles will limit the types of organisms that evolve, but there are many body plans that fit the physical requirements. Evolution can go from less complex to more complex organisms, but it can also go the other way—towards simpler organisms, including parasites; this is 'reductive evolution'.

19. Humans are 'a part of nature', 'not apart from nature'.

Nature was not created *for* humans, nor are we exempt from the laws of nature. This is one of the most far-reaching conclusions from evolutionary theory. Recall some of the aspects of indigenous knowledge from the first chapter. The idea that plants and animals were created by some deity 'for humans' was one of the most widely believed throughout the world.

For many people, a difficulty with a full evolutionary theory is that it did not appear to explain or at least demonstrate a need for ethical and moral systems that characterize human (and many animal) societies. Several scientists accepted evolutionary ideas but still sought a non-evolutionary explanation of human ethical, moral, and religious systems. This is one of the active areas of current research with studies on higher primates (especially apes), human development, and computer simulation. The origin of language is now seen as firmly embedded within the great apes, but what about religion? Is there an advantage to individuals if they are members of a group with firm shared beliefs? How is it that early religions appear to encourage killing of members of 'other' groups? Is this the same basic phenomenon that we see in some chimpanzee societies where they are tolerant and supportive to others within the group but willing to kill (male) chimps from other groups? Is there an evolutionary advantage to individuals that share strong beliefs? In addition to the positive side of religion, is there a 'dark side'?

All the differences we see between human and chimpanzee genomes are normal microevolutionary processes of genetics: point mutations, small insertions and deletions, duplications of parts of the chromosome, a few new enzymes, activation of retrotransposable elements, and fusion of two chromosomes to form chromosome 2 in humans, etc. (see Chapter 5). All perfectly normal, there is nothing special about humans as far as our genetics is concerned. There are certainly extremely large numbers of point mutations and small insertions and deletions, but again the numbers are as expected from measuring natural variability in populations. In contrast, wouldn't it be fascinating if some kindly creator, or group of itinerant space travelers, had inserted into the human genome a whole lot of genes for wisdom and intelligence? Just think: genes for wisdom and intelligence—all we would have to do is find some way of turning those genes on!

20. Additional mechanisms contributing to evolution will be found.

This is really self-evident, but it is important to state it for several reasons. From a logic viewpoint, it is necessary to distinguish between using only mechanisms that can be studied in the present—from assuming that we know all the mechanisms. We have seen a huge increase in knowledge during the twentieth century, which continues today. There are major gains in molecular biology, genetics, biochemistry, physiology, development, ecology, and behavior; and we would be very surprised indeed not to expect similar gains this century. As just one example, we still do not know all the (chemical?) mechanisms required for the origin of life (yet probably most scientists think that such mechanisms exist).

A limitation is that any such new mechanism must be able to be studied in the present ('causes now in operation'), including impacts. Darwin considered sexual selection to be different from natural selection though modern work (by concentrating on the number of viable offspring rather than survival of the parents) and usually treated sexual selection as an aspect of natural selection. In the next chapter, we will consider some of the new developments from the twentieth to twenty-first century.

What the theory did not explain was the mechanism for the continued generation of variability. Darwin (and many others) noted that in many plants, sexual reproduction led to more variability among offspring than did nonsexual (asexual) reproduction. This observation by itself still does not 'explain' the diversity, but it did mean that, in principle, mechanisms that continue to generate diversity must exist. It was not until the next century, after the rediscovery of both Mendel's genetics and of DNA, that the problem of the origin of variation could be tackled. There is still the basic problem of the linkage between inheritance (genetics) and form (morphology) of an organism.

But while the concept of competition was combined with the assumption that all individuals in a species are 'essentially' the same, then all members were treated as equivalent, and the intra-specific aspects was not recognized. What had been accepted earlier was that selection could remove some less-fit individuals that deviated from the perfect type or essence of the species. The intra-specific variants most different from the original type would be lost, thus maintaining the species 'essentially' unchanged. Consequently, even when selection had been recognized, the original concept of species inhibited thinking about continued evolutionary change from the original type.

Summary

There are other ways to analyze Darwin's theory. Mayr (1985) considered it as five theories which cover much the same ground, and the larger categories in the present analysis are similar to those of Mayr. What is important here is that the overall theory, together with its consequences, is relatively complex even though each component (1–20) is relatively straight-forward. He considered all the steps as being one reason why his thinking and reasoning were not equaled or surpassed until the 1930s. The questions now being asked about the sufficiency of evolutionary theory form the 'third Darwinian revolution', the first two being the acceptance that evolution had occurred (for biologists, this mostly in the 1860s and 1870s) and the second being the necessity of microevolution (mostly by the 1940s and 1950s).

Every few years, a would-be scientist comes up with a new 'non-Darwinian' theory of evolution. Unfortunately, these usually have at least one of two features:

> They have 'rediscovered' for themselves something well known
> to Darwin and earlier workers (for example, King and Jukes
> 1969; Eldredge and Gould 1972).

> They do not define 'non-Darwinian'. Which of the twenty points
> listed above do they disagree with, or are they rejecting all
> twenty of them? (There are 1,048,575 possibilities ($2^{20}-1$).
> What hypothesis are we to test?)

For example, King and Jukes (1969) suggested 'neutralism' as 'non-Darwinian', but Charles Darwin was there before. His statement from Chapter 4 of *Origin*—'*Variations neither useful nor injurious would not be affected by natural selection, and would be left either a fluctuating element, as perhaps we see in certain polymorphic species, or would ultimately become fixed*'—covers the main points. These were the two points (some mutations being neutral and the fixation of these) that neutralism advocated. And then Eldredge and Gould (1972) suggested unequal rates, but again Darwin had already had a heading in his Chapter 10: '*On their different rates of change*' (p. 312). And he had said, '*Species of different genera and classes have not changed at the same rate, or in the same degree*' (p. 313). And as to reading non-Darwinian proposals, it has often been impossible to tell exactly what the authors are criticizing. Are they are just unthinkingly against 'neo-Darwinism'? Yes, we must make progress, and there will be additional things that certainly Darwin did not know about—but that is not the issue here. Some people think there are some errors. So let's test them, but it does need that the potential errors be specified scientifically. The real issue is 'causes now in operation'.

Was Charles Darwin 'perfect' as a thinker? No way! He also tried to think up a theory of the transmission of characters called pangenesis, and it was wrong! (He talked about his 'much maligned theory of pangenesis'.) But perhaps the main point is that he assumed testable mechanisms, a principle he adopted from geologists James Hutton and Charles Lyell. All of the twenty points are important; this is the real issue.

CHAPTER 4

Newer Developments in Evolution

Our knowledge is never static—we are always learning about new areas and new ideas. And as we learn about new things, newer areas develop, and our aim is to consider some of them in relation to evolution. So how has Darwinian evolution stood up with new knowledge? We always want to know this in any Popperian analysis; it is always a sensible question, and we do expect that many ideas will change/evolve through time. But it does appear that Darwin's view of evolution has stood up well even though there have been many new developments. I suggest that it is because Darwin concentrated largely on known mechanisms that operate on the present and that, as many new things have been learned, they still all operate today.

Yes, there was one thing that Charles Darwin did get wrong, but that was largely recognized in his time. As we have indicated, he did suggest that his 'much abused' theory of pangenesis (inheritance) was wrong. And we will not consider the uses, and abuses, his theory was put to in the social arena: 'social Darwinism'. This is better called 'Spencerism' (after Herbert Spencer) and is quite a misnomer for Charles Darwin's ideas even though it is often (and erroneously) attributed to Darwin. This is a very conservative application of evolution to society, often being used to justify (to our way of thinking) a harsh interpretation of evolutionary theory to society—competition was 'natural'; therefore society should follow evolution. However, cooperation between individuals (or species) is also favored by evolution!

But before we begin this section, we should quickly mention the structure of the three main classes of macromolecules: proteins, RNA (ribonucleic acid), and DNA (deoxyribonucleic acid). Proteins are made up from twenty amino acids that all have the same structure of a carboxylic acid, an amino group on the next carbon atom, plus a side chain. It is the side chains that differ between the twenty amino acids, and many (most) can be catalytic in the correct circumstances. They are coded for by non-overlapping triplets of messenger RNA (mRNA), and the proteins are now the main catalysts of the cell (the enzymes). From the sequence or the mRNA, it is relatively easy to read off the protein sequence, though to be active, the protein has to fold—and the principles are still being established.

There are the two nucleic acids: RNA and DNA. Each is composed of nucleotides, a sugar, and phosphate. For RNA, cytosine pairs with guanine (C=G, with three hydrogen bonds) and uracil with thymine (A=T, with two hydrogen bonds). RNA has the fully oxygenated sugar (a five-carbon sugar ribose), and DNA has a sugar derived from ribose, namely deoxyribose. However, RNA is largely an unpaired molecule but does fold into tertiary structures. We still have CG pairing in DNA, but now we have adenine pairing with thymine, again with two hydrogen bonds. The difference between ribose and deoxyribose is that ribose has one more OH group than 'deoxy'-ribose, which has an H attached to the second (2') carbon in the ring. It also has larger grooves, which makes it easier to be processed by enzymes. RNA is usually single stranded (and folds into many shapes), and DNA is double stranded. DNA, being double stranded, is copied much more accurately than RNA—the new strand can be compared against the old strand. Now we will get back to learning about inheritance.

The Rise of Genetics

The mechanisms of heredity were still a major problem in the late nineteenth century, but why were the ideas and results of Gregor Mendel of the Czech Republic (1822–1884) not appreciated at the time? As we have already mentioned, one idea from Sandler and Sandler (1985) is that Mendel did not attempt to explain 'development' (how the characters were actually formed). The Sandlers' suggestion is at that time, the two ideas (genetics and development) were linked together— initially considered as basically the same problem or the same question. It was only after the work of German August Weismann (1834–1914) that the two questions were accepted as being separate. Thus Weismann clarified the problem by separating

> the problem of inheritance from
> the problem of development of the characters.

Mendel also concluded that there were separate germ line and somatic tissues. Thus he had established much of the genetics (inheritance) aspect, but not how the characters actually formed—that is being left to 'developmental biology'. So fortunately, by the year 1900, the problems of genetics (inheritance) and of development were generally recognized as being separate issues. Developmental biology is still 'developing' (although it certainly has made great strides), but we see it today as being a separate question. To a large extent, this (according to the Sandlers) allowed the rediscovery/acceptance of Mendel's laws of inheritance in 1900.

Once the two problems (hereditary and the development of characters) were accepted as separate issues, it was possible to recognize (rediscover) Mendel's contribution—he had solved much of the inheritance part. It is only in the last fifty years or so that it has been increasingly possible to study effectively the genetic and biochemical basis of development. In other words, the problem of inheritance had to be more or less established before the problem of development could be studied effectively. It was

Weismann who separated these two concepts even though he was wrong on many details—but that is good progress. One example was that he assumed the inheritance of acquired characters was impossible because of the early separation in development of the germ (reproductive) cells from the rest of the body (somatic cells). This might be true in many animals, but not in plants, fungi, and most other organisms. Okay, many plant cells do not usually contribute to the next generation, such as most root cells. Thus Weismann's conclusion (separating inheritance and development) was correct, but his explanation was not, and he did not include epigenetics. But more importantly, his conclusions allowed the rise of genetics.

So the first major new development in understanding evolutionary theory was this rise of genetics. Gregor Mendel is well known to have discovered some regularities (laws) of inheritance, but most of his experiments were carried out on the pea plant (*Pisum sativum*), though there was a long history of studying many aspects of inheritance (what we now call genetics) on domesticated plants and animals in Brno (Poczai et al. 2014). In the year 1900, rediscovery by at least two researchers (Hugo de Vries and Carl Correns and possibly by Erich von Tschermak) is recognized as a major step forward. Mendel's results covered mainly the seven characters of seed shape, flower color, seed coat tint, pod shape, unripe pod color, flower location, and plant height. His conclusions were strongly reinforced because of a lack of linkage—there were seven chromosomes, and Mendel had seven characters that showed the 3:1 ratio, one on each chromosome.

However, the concept of chromosomal linkage was quickly established and helped show that the chromosomes (and their molecules) were important for inheritance. This helped identify the chromosomes as the carriers of genetic information. There are several books available that describe aspects of the discovery of modern genetics and the rise of neo-Darwinism that integrated Darwinism with the new study of genetics (e.g. Mayr and Provine 1980). Perhaps the main point here is that the discovery of genetics eventually (and the emphasis on the eventually) supported the twenty points laid out in the previous chapter—the Lyell/ Darwinian approach of studying mechanisms that can be studied in the present. In the early twentieth century, there had been a tendency to ignore/neglect some of them (see later). Anyway, we now fully accept the discoveries of genetics to be essential, and now we want to go much further.

The Modern Synthesis

Early in the twentieth century, there were apparently at least two main streams of evolutionary thought: the 'mutationists' and the 'Darwinians'. The mutationists tended to be laboratory-based geneticists who studied larger and discrete mutations that were easily counted in the lab, but who thought natural selection less important. At one extreme, some people thought that large mutations might even form new species—'all at once'! On the other hand, field workers—who studied natural populations, but not inheritance—argued that the important continuous characters (such as height) did not show Mendelian inheritance. Therefore, Mendelian genetics was supposedly not important 'in the field'. So initially there was some disagreement between the geneticists and other biologists, and everyone has their own theory of what was missing.

The point is illustrated by two extracts, the first from 1921 and the second from 1929.

There is a strong tendency in these days to admit natural selection only as a 'merely negative force' and as such it has even been dismissed as a truism. It may be that the theory of natural selection as Darwin and Wallace understood it may some day come into its own again: but our present total ignorance of variation and doubt as to other means of change we can form no clear idea of the material on which selection has to work and we must let the question rest. For the moment, at all events, the Darwinian period has passed. We can no longer enjoy the comfortable assurance which once satisfied so many of us that the main problem had been solved. All again is in the melting pot. By now a new generation has grown up that knows not Darwin (D. H. Scott 1921, presidential address, botany section of the British Association).

Whilst the fact of evolution is accepted by every biologist the mode in which it has occurred and the mechanism by which it has been brought about are still disputable. The only two 'theories of evolution' which have gained any general currency, those of Lamarck and Darwin, rest on a most insecure basis; the validity of the assumptions on which they rest has seldom been examined and they do not interest most of the younger zoologists.
(D. M. S. Watson, presidential address, zoology section, British Association for the Advancement of Science).

'*Knows not Darwin*' and '*do not interest most of the younger zoologists*' are powerful statements indeed! These two quotes illustrate that by the early twentieth century, Darwin had convinced biologists (and many others) that evolution had occurred (macroevolution in our terminology), but he had not convinced some people that the processes of natural selection were either necessary nor sufficient for evolution. The low point of the acceptance of his mechanism was in the 1920s, but then the situation changed quite rapidly.

Sergei Chetverikov (1880–1959) in Russia and later both Fisher and Haldane (in England) and Sewall Wright (in the United States) showed that the two schools could be combined. So from the mid-1920s onwards, the developments in population genetics led to the synthesis of these two streams of biology: the laboratory geneticists and the field biologists. Early work occurred in Russia (before Stalin became powerful). Chetverikov undertook both theoretical and experimental work with populations. These studies were stamped out in Russia following the rise of Stalin (because the work did not conform to the then Marxist ideology of the inheritance of acquired characters, favored by Lysenko). However, this early work was influential among some Western scientists, particularly from expatriate Russians and from scientists who had visited the Russian laboratories.

In Europe and North America, Fisher (encouraged by Charles Darwin's grandson Leonard Darwin), Haldane, and Sewall Wright did theoretical (mathematical) work that led to a synthesis of evolution and Mendelian genetics. This was called variously 'the synthetic theory', the 'new synthesis', and 'neo-Darwinism'. These workers were fairly mathematical and showed quantitatively that the inheritance of the discrete genes as proposed by Mendel could give the continual change of qualitative characters studied by naturalists in the field. The ability to give a mathematical precision greatly increased the explanatory power of theories even if many of the conclusions were statistical in nature (which of course fitted Darwinian probabilistic theory very well; see the point 14 of Chapter 3). Indeed, population genetics was important in the development of statistical thinking in that many new methods were developed from problems in genetics. Nowadays statistics is very broad and includes many other aspects and certainly not just genetics.

Many other authors in the 1940s and 1950s then explained the synthesis to other biologists (that is, they left out the much of the mathematics!). Of these, Julian Huxley's *Evolution: The Modern Synthesis* (1942) is a good overview of neo-Darwinism and should always be consulted by anyone interested in classical neo-Darwinism. Several misunderstandings of what neo-Darwinian theory was about arise simply because the critics have apparently not read Huxley's book! As we have said, the book edited by Mayr and Provine (1980) has many articles covering the development of the synthesis, but the details are outside our scope here.

From our viewpoint today, we might criticize the original synthesis as still being too narrow (for example, it left out most of epigenetics) and still was too deterministic. However, the subject, like any area of science, has continued to develop with perhaps the important gains being in the area of evolutionary ecology and behavior (in a broad sense, that includes plasticity in plants), molecular evolution, and mathematical precision. But related to these was the conclusion that the chromosomes were the centers of inherence. Thus it was important next to determine the structure of the molecules in the chromosomes; they needed to be identified as the carriers of genetic information. So it was very important for the next step to identify the chromosomes (and the nucleus) as carriers of genetic information.

The Nucleus and DNA as Carriers of Genetic Information

The next step was determining that the nucleus (in eukaryotes that have a true nucleus) and the chromosomes were the carriers of genetic information. An additional step then was that DNA, as a component of the chromosomes, was the carrier of the genetic information. Perhaps it was a bit easier in eukaryotes with a nucleus to soon show it carried the information. One approach that showed the nucleus was the carrier of inheritable information is the following: There are some very large cells of green algae (*Acetabularia*) where the large single nucleus has DNA duplicated many times. Nuclei were transferred from a first species of *Acetabularia* to a second species, and the tops of the cells (at the other end of the cell from the new nucleus) were transformed into the growth form of the new nucleus! The simplest interpretation was that the nuclei carried the genetic information for the cells; they were the only part of the cell transferred between the species. There are other experiments that support this as well, but that one was quite crucial.

There were quite a few experiments that then showed that chromosomes carried genetic information, and some were even earlier than those reported above on nuclei. One of the first studied were deletion mutations observed in some chromosomes formed during insect (*Drosophila*) development. At a certain stage of development, the DNA was duplicated in each chromosome so that the structure of the chromosomes was visible in a light microscope. Thus these chromosomes allowed visualization, and each chromosome could be identified—an important point. Some of the absent markers were quickly shown to correspond to deletions of a particular gene and thereby showed that the chromosomes themselves were carriers of the genetic information.

The next step probably was to show that the chromosomes (in prokaryotes in particular) contained DNA (discovered by Johannes Friedrich Miescher (1844–1895) from pus cells in 1861 in the kitchen of Tübingen castle—okay, it was then a laboratory for the local university). But the DNA was identified before the 1950s as containing the genetic material.

DNA, Protein, and RNA

A major gain in knowledge was learning the structure of DNA, proteins, and RNA. First, DNA had to be identified as the carrier of genetic information, and this was known by the early 1950s. As is now well known, James Watson and Francis Crick (1953) determined the structure of DNA by X-ray crystallography and demonstrated the binding abilities of the nucleotides—it was relatively easy to see how DNA was replicated. The structure Watson and Crick reported had cytosine pairing with guanine (forming three hydrogen bonds, shown as $C{\equiv}G$) and then adenine pairing with thymine (with two hydrogen bonds, shown as $A{=}T$). This certainly gave a method that, in principle, allowed the replication of the DNA; adenine could pair with thymine and cytosine with guanine. There is a process of checking that the new strand has the right pairing with the old strand—the 'new' and the 'old' strand can readily be identified. The overall mechanism allows (in some cases) about one error in 10^{-11} or 10^{-12} nucleotides added! Brilliant accuracy—and it gives a reason why DNA replaced (the basically single-stranded) RNA as the genetic material (see the Eigen limit later). There is now a sizeable 'industry' in determining the rates of mutations. But the copying accuracy is brilliant. In addition, there are some small insertions and/or deletions and sometimes duplications of whole genes. We will come back to this later in the question of the origin of life.

But perhaps of even more interest was how DNA might be controlled. But all this also was helped by the identification of chromosomes and then of DNA as the carriers of genetic information. For this aspect, we consider the work of François Jacob and Jacques Monod (1961) on the *lac* operon of the bacterium *Escherichia coli*; this turned the set of genes on or off. When I was a graduate student, Jacob and Monod (of France) were our heroes. Watson and Crick had determined the structure of DNA (someone had to do it was our reasoning), but Jacob and Monod determined how DNA really worked—a much greater achievement (so we thought). Nowadays it has reversed; everyone knows about Watson and Crick, but hardly anybody (among our undergraduates at least) knows of the work of Jacob and Monod and their work on understanding the 'operon' structure of prokaryote (akaryote) genes. Perhaps the main point was that the Jacob/Monod model had messenger RNA (mRNA) between DNA and protein, and this is still accepted today. The *lac* model also had other genes involved whether or not the mRNA is 'turned on' and expressed as mRNA and then as protein.

It also had to be determined that proteins had both a definite sequence of amino acids and that the proteins had a definite three-dimensional structure. Proteins are made up of long chains of amino acids and fold into three-dimensional structures—usually automatically, though sometimes with the aid of special folding proteins that assist their folding. It was a surprise to some chemists that the proteins had a defined sequence, and a definite three-dimensional structure. So first we should give some of the early history. In his book *The Logic of Life*, François Jacob (whom we met earlier for his work discovering operons) covers some of the very early work in genetics. He explains some of eighteenth-century concerns of 'inheritance'. Buffon, for example, could not understand how three-dimensional information was passed onto the next generation, so he assumed that (somehow) three-dimensional information from, for example, the structure of the heart had to be passed on between generations of animals. This (in Buffon's understanding from the eighteenth century) was that little 'molds' (the *interior milieu*) were necessarily passed on between generations—for example, giving the shape of the heart, the kidneys, and so on. Nowadays we just take it for granted that linear information (for example, the linear information in RNA and DNA and proteins) is sufficient. The laws of chemistry and physics do not have to be passed on between generations (they are in a sense constants that are always present). So this was a great advance; the laws of chemistry and physics were always present—linear information was sufficient. This was an amazing discovery from the twentieth century.

So some early experiments determined the structure of proteins and found that they normally had both a defined sequence and defined shape, as shown by X-ray crystallography. This was very informative at the time, though nowadays it is 'common knowledge' and somewhat taken for granted—it is now a thoroughly evaluated idea. As we have said, there had been some initial uncertainty about whether proteins (as enzymes) had defined structures and shapes—they certainly do, though the structures can be flexible and modified by changes to the enzymes, such as phosphorylation. Currently larger and larger protein structures (sometimes with RNA, such as the ribosome) are being solved, initially by X-ray crystallography. However, there is now considerable interest in both NMR (nuclear-magnetic resonance) and cryo-electron microscopy. Larger and larger structures are being solved; one of the latest is the structure of the nuclear pore complex. However, we still do not know the structure of the cell as a unit, and we need to be able to more or less 'predict' three-dimensional information, especially how the structures may carry out catalysis (if they do).

Molecular crowding is another more recent understanding. The distance between proteins is quite low, and this means that diffusion of proteins (because they are macromolecules) will be relatively slow—there is not much space between the proteins. Conversely, though, the smaller molecules may be more concentrated and thus accelerate the rate of chemical reactions. This means that we are getting somewhere desirable, but we still cannot yet say what a particular protein might do. So although there has been good progress in understanding inheritance, we do need to go further—much further. This still seems that we are a long way from predicting what effect a given mutation is likely to have on a phenotype (including resistance to disease). Will a specific mutation be neutral, advantageous, or disadvantageous? And under what conditions? A small part of the problem is that the many small RNA molecules affect the level of expression in different cell types, especially in multicellular plants and animals. However, the overall problem is beyond our current understanding (but we will get there—we must).

Epigenetics

This is a very interesting development that is certainly ongoing, and it is broadening the scope of 'inheritance', particularly in eukaryotes (cells with a nucleus). The first point is that classical genetics traditionally considered inheritance to be limited to changes (mutations) and to the DNA sequence and that any changes to the DNA sequence (at least in germ line cells in multicellular organisms) can be passed on to future generations. However, genes are expressed at very different levels in the tissues of multicellular organisms, and even predicting gene regulation in unicells is difficult when cells may be in quite different environments.

Let's say what 'epigenetics' is first because it is sometimes considered 'Lamarckian' (Jean-Baptiste Pierre Antoine de Monet, Chevalier de Lamarck (1744–1829)). However, this is a misnomer; there appears to be no evidence that Lamarck proposed the inheritance of acquired characters as opposed to the general assumption that that was the mode of inheritance. There is no discussion in his book (Lamarck 1809) about the mode of inheritance. However, he was an early evolutionist, though his understanding of it was weak by modern standards.

But first, there has to be linkage to the germ line; any changes outside the germ line will not be inherited, though this may not be a problem for many unicellular eukaryotes. Perhaps the most important issue is that we still do not know why (or really how) the different cells in multicellular plants and animals express genes differently. After all, they have the same gene content. The existence of epigenetics was rather controversial initially (because it looked like the inheritance of acquired characters), but it is now accepted as part of inheritance.

It is now increasingly accepted that some other changes (in addition to normal mutations) may also be passed on to future generations. For example, eukaryotes have several (five) histone proteins attached to their DNA in pairs, and these form regular structures along the DNA (the nucleosome), and the tightness of theses nucleosomes affects gene expression. An important point is that methylation, phosphorylation, etc., of DNA and of these histones are part of normal evolution (at least in eukaryotes that have the nucleosome structure). Covalent modifications of either DNA, for example, cytosine methylation and hydroxymethylation or of histone proteins (e.g. lysine acetylation, lysine and arginine methylation, adenosine methylation, serine and threonine phosphorylation, and lysine ubiquitination and sumoylation (the latter two are modifications by proteins)) play important roles in many types of epigenetic inheritance. In addition, there are many small RNA molecules (generally called microRNAs) that affect the level of protein expression.

An initial way to modify gene expression is the post-translational modification of the amino acids that make up these histone proteins. If the amino acids that are in the chain are changed, the shape of the histone might be modified. It is possible then that the modified histones may be carried into each new copy of the DNA. Once there, these histones may act as templates, initiating the surrounding new histones to be shaped in the new manner. By altering the shape of the histones around them, these modified histones would ensure that a lineage-specific transcription program is maintained after cell division.

A second way is the addition of methyl groups to the DNA, mostly at CpG sites (cytosine followed by guanine), to convert cytosine to 5-methylcytosine. This performs much like a regular cytosine, pairing with a guanine in double-stranded DNA. However, some areas of the genome are methylated more heavily than others, and highly methylated areas tend to be less transcriptionally active through a mechanism not fully understood. Methylation of cytosines can also persist from the germ line of one of the parents into the zygote, marking the chromosome as being inherited from one parent or the other. The genome structure of eukaryotes is most interesting.

There is now good evidence for several aspects of epigenetics in all the main groups of eukaryotes (Penny et al. 2014), but we do need some additional information to increase our confidence here. There are quite a number of processes included under 'epigenetics'. For example, it includes methylation (of both DNA and histones), acetylation, and phosphorylation. One of the main uncertainties has been an ancient group of eukaryotes (Excavates) as to whether they had epigenetics, but some aspects (especially acetylation) certainly appear to occur in Excavata, such as within *Giardia*. Histone lysine acetylation also plays an essential role in regulating the life cycle from cyst to trophozoite. Acetylation has also been characterized in *Plasmodium* and *Entamoeba*. Other protists such as *Toxoplasma* have a wide repertoire of histone modification proteins. Enzymes used for histone arginine methylation (especially protein arginine methyltransferases, PRMTs) can be found in diverse protist species (though not all have been demonstrated to be functional). DNA methylation is known in diverse eukaryotes from mammals to nematodes to ciliates, where methylation and hydroxymethylation are involved in DNA elimination. However, it is still not known whether the mechanism characterized in the ciliate *Oxytricha trifallax* is descended from the mechanism for example in nematodes. The enzyme DNA methyltransferase (dnmt) apparently has no homolog in *O. trifallax*, but DNA methylation inhibiters do disrupt the process, somewhat indicating that the process works in a similar way.

There is also a strong link between epigenetic mechanisms and RNA (reviewed in Collins et al. 2010). Even mechanisms that appear to have evolved within animals (X-chromosome inactivation (Xist), see Collins et al. 2010) or plants (siRNA viral defense that uses RNA-directed DNA methylation) connect RNA processing to epigenetic marking and unmarking. Such RNA-directed modifications occur in other eukaryotes, although the mechanisms are less understood. In *Plasmodium*, long ncRNAs are involved in a mechanism similar to Xist. So our current conclusion is that epigenetics was also in the LECA (last eukaryotic common ancestor), though more information will certainly be helpful here.

So we do see epigenetic mechanisms reported throughout eukaryotes at least. However, with our knowledge still growing in protist systems, it is still too early to say whether all epigenetic mechanisms stem from an ancestral state. So at the moment, it appears as if most of the epigenetic mechanisms occurred in LECA—the ancestor of the crown group of eukaryotes. This is an important problem for the future. We need to know if epigenetics is universal in eukaryotes at least.

The RNA World

In Chapter 1, we saw that we no longer accept 'continued spontaneous generation'. But life must have started naturally in the first place. It is most important that the whole question of the origin of life is considered a sequential process and definitely not a one-step process—nobody appears to assume that life arose instantaneously, all at once. However, proteins currently do all the work (carry out the reactions), and DNA has all the information (but doesn't carry out reactions on its own)! So which came first—the proteins that carry out the reactions or the DNA that does 'nothing' (but stores all the information)? This is a classic 'chicken and egg' problem! We have to have chickens to lay eggs, but we have to have eggs in order to get chickens. The answer to the 'chicken and egg' problem' is the same as the 'DNA and protein' question—neither! The standard answer (for proteins and DNA) is that RNA came before proteins or DNA—'ribonucleic acid' (reviewed in Robertson and Joyce 2012). These macromolecules are similar to DNA in that they are mostly composed of the same nucleotides (but have uracil instead of cytosine) and have ribose instead of deoxyribose as the five-carbon sugar.

RNA can both code for proteins but also carry out catalytic reactions by itself. It was an important discovery when researchers found 'ribozymes'—RNA molecules that carried out reactions. They were named from a combination of 'ribo' from RNA and 'enzymes' (from proteins), hence the term 'ribozymes'. Some ribozymes just bind a particular molecule. Thus RNA can both code for proteins (such as in mRNA), and it can carry out reactions (it is normally single stranded and can fold with the hydrogen bonds between some nucleotide bases). There are also viruses (such as the 'flu virus) that are composed of only RNA and proteins; they lack DNA altogether. Okay, it was later found that DNA also could be catalytic, but not to the same extent RNA is, and there appears to be no naturally occurring deoxyribozymes (whereas there are thousands of ribozymes).

Anyway, there is now good evidence for the three-step scenario that

> RNA was the first macromolecule; then
> RNA made peptides, then proteins, and then later
> proteins invented DNA (with its much more accurate copying mechanism).

This is the classic 'RNA-world' scenario; RNA $\rightarrow$ coded protein syntheses $\rightarrow$ DNA (but it is still possible that some other polymer came before RNA). There are at least two versions of the RNA world and a continuum between them. The first is with many amino acids (and short non-coded peptides) and one with just pure RNA. The former is often called the RNA-peptide world. I guess that we generally prefer the version with peptides, including dipeptides, tripeptides, etc. It appears impossible to generate proteins unless amino acids, dipeptides, etc., are involved and dipeptides, for example, are catalytically active. Evolution has no foresight; it can't make peptides because 'they will be useful sometime (millions of years?) in the future'. They have to be useful here and now—in the present! So generally we consider an 'RNA-only world' as a misinterpretation—a misunderstanding of the RNA world. That is, we currently favor an RNA-peptide world. Anyway, it should be obvious that there are many intermediates on the way to living systems, and this is recognized by all realistic assumptions.

Before we go any further, we should point out that it is virtually established that there is a single origin of life in this planet. The probability of having the same amino acids, the same code, and the same macromolecules (RNA, proteins, and DNA) is vanishingly small. This conclusion is quite independent of whether life arose here or came from some other part of the universe—panspermia. So we do not know whether 'life' on another planet uses the same amino acids and has the same code—an interesting question. All life on planet Earth appears to have arisen only once. Do we 'believe' this (and not test it), or do we treat it as the best hypothesis that we currently have?

The Eigen Limit

But first, we should tell you about what we call the 'Eigen limit'—the number of errors the length of a coding sequence that can be maintained by natural selection for a given error rate. We sort of joke a little about Manfred Eigen. He won a Nobel Prize in Chemistry when he was about 40 (for studying ultrafast reactions by spectroscopy). 'Chemistry is easy—now I will study biology' is the joke. No, chemistry is not easy, but we are pleased that he turned his attention to biology. He showed relatively quickly that there was a relation between the error rate (per nucleotide) and the length of the sequence that could be maintained. For example, he assumed a 'master sequence' that was copied optimally/fastest. And he allowed a slight reduction for every mutation—slight in that it might be only a 1% reduction in rate for every mutation. These are only macromolecules, and so we certainly cannot ascribe them any intelligence! A slight reduction in copying rate is a semirealistic assumption. So there are three things:

the length of the sequence being copied
the error rate per nucleotide added
the reduction in the rate of copying per mutation (see Figure 4.1)

Under these conditions, there becomes a relatively stable distribution of original sequences, sequences with no mutations, sequences with just one mutation, sequences with two mutations, sequences with three mutations, etc. (Figure 4.2).

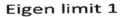

master sequence

Eigen limit 1

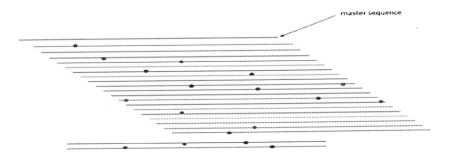

Fig. 4.1. The Eigen limit. The master sequence is the one copied optimally, with those with one error, two errors, etc., copied less well, say, 1% less for every error. For this calculation, we combine all the sequences with one error, two errors, three errors, etc.

Copying RNA has a much higher error rate than copying DNA (the sequence of the new strand of DNA can be compared against the old strand, and any errors can be corrected). In agreement with the Eigen calculations, RNA viruses have a much shorter sequence length than DNA viruses and DNA-based organisms (that is, all organisms—we do not know of any RNA-based organisms, but there must have been some in the very early days). Basically, the Eigen calculation is an important contribution, and in practice, RNA ribozymes are known to successfully copy RNA molecules.

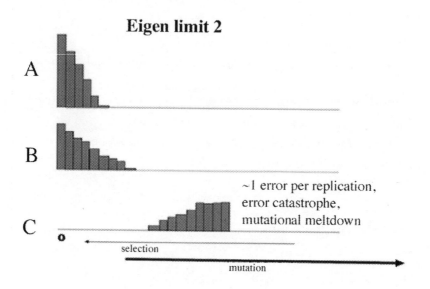

Fig. 4.2. Under these circumstances, there is a possible stable relationship between the number of errors and the mutation rate, and at lower error rates, the master sequence is maintained (A with fewer mutations, B a little more, and C more still). As the mutation rate increases (for a fixed length) we eventually go into 'error catastrophe' (or 'mutational meltdown'), and the sequence can no longer be maintained—it becomes full of errors.

The Origin of Life

The first issues to consider is the nature of living systems and to ask whether life is a natural property of matter (it was certainly considered to be earlier when it was assumed that life arose spontaneously—a different question). Harold Morowitz (Skoultchi and Morowitz 1964) carried out a particularly informative experiment about the nature of living systems—if I had to award a top prize for twentieth-century experiments on 'life', I would have to award it to him. For example, he took an *Artemia* (a brine shrimp) embryo down to 2°Absolute (~–271°K), kept it at that very low temperature for six days, and slowly warmed it up—and it survived, grew and, reproduced. At that very low temperature, all the electrons would have been in their 'ground state', their lowest energy level. So the conclusion was that life appears to be a natural property of matter—place the right chemicals in the right position and the resulting cell will be alive! I like to call this *coito ergo sum*. (Okay, this misquotes Descartes a little. He said *cogito ergo sum*, '*I think, therefore I am*'. But I prefer to drop the *g*. This version is translated politely as '*I reproduce, therefore I am*'.) Good biology!

It is probably important that these *Artemia* embryos are used to being dehydrated, though an even earlier experiment was carried out with the bacterium *E. coli*, where it also survived. However, it was very interesting to see the experiment repeated with an organism (*Artemia* embryo) that has a heart, nerves, brain, etc. Was it just a fluke that they survived? Probably not, but it would certainly be worth repeating the experiment with other organisms that have the ability to survive desiccation—for example, the 'resurrection grass', including some members of the grass genera *Sporobolus* and *Oropetium*. It does appear that several grass species have the ability to survive drying, and it would be good to see whether some of them have the ability to survive at 2°A. Is it truly a general property of life? It does seem so. It works with both *E. coli* and with *Artemia* embryos. It probably also works with tardigrades, which are also animals.

Morowitz's hypothetical counter-example was a jellyfish (which has a circular neuron). If one part of the nerve is touched, a spike of activity would go around in both directions via the circular nerve, and then when the two signals met on the other side, they would cancel each other out. If, however, some unkind physiologist held a piece of ice briefly on one side of the stimulus, it would stop that side of the nerve reaction. Quickly taking away the ice away would leave the stimulus going one way only—it obviously did not cancel out when it reached the other side. The stimulus just went around and around, seemingly forever.

However, if the jellyfish could be taken down to 2°A, then that signal would be lost—it depends on some of the electrons being at a higher energy level; so when all the electrons are the minimum level, the signal just disappears. (The experiment is hypothetical in that the jellyfish would not survive being frozen, but that is a separate issue, and it doesn't matter for illustrating the principles; the *Artemia* embryo survived.) But because the very low temperature worked for both *E. coli* and the brine shrimp, the conclusion of Morowitz was that that 'life is a natural property of matter'—if you got all the chemicals in the right arrangement, then it would be 'alive'. I guess we imagined a myriad of chemists tapping away at their typewriters (in the 1960s model, computers in the modern world), making chemicals for testing. Nevertheless, it is a major conclusion, but we still do not yet have the 'right arrangement'. So yes, it does not seem anything chemically unusual about 'life'. We will see later that some good progress has been made in studying the origin of life.

In a quite separate experiment, Craig Venter synthesized (by steps) a DNA genome and transferred it step by step to a bacterium. Eventually, he had replaced all of the DNA with the synthetic copy; he synthesized, say, 1,000 nucleotides at a time and integrated them with the existing genome. So the 'genome' was in that sense all synthetic, and it still worked. There was nothing 'special' about synthesized molecules in that experiment. Now he has produced an 'artificial cell' with a reduced number of genes. So yes, there is good progress in the steps towards the origin of life, and it does seem that life (based on taking living systems to 2°A) is a natural property of matter.

Similarly, the origin of life had to be one of many molecules working together in an interactive system. It would not really help if 'the system' only made (or had available) ten of the twenty amino acids or three of the four nucleotides! The system has to make all of the amino acids and all of the nucleotides in order to be able to work properly—a fully cooperative system. There has been a tendency to see the negative side of evolution, emphasizing the 'selfish' aspect of evolution. (On a separate issue, this is evidence that all life is homologous; all life is based on the same principles and the same code—it did not have to be that way.)

Unfortunately, there is no money to be made in solving the question of the 'origin of life'—so it is not really a major area of study! However, there are a few hardy chemists who study the issue. However, there are probably hundreds of intermediate steps towards an origin of life— nobody seems to assume that life arose 'all at once' by some miracle.

Higher or Lower Temperatures to Start Life

Did life start at higher or lower temperatures? By higher temperatures, we mean above about 80°C, and this includes black smokers at the bottom of the ocean. Lower temperatures are from frozen ice to about 20°C. Clearly there is a large intermediate zone of intermediate temperatures from 20°C to 80°C. So the main question here is at higher temperatures, we are really dealing with the origin of life as a kinetic (enzymatic) problem, and this is presumably better at higher temperatures. Or is the origin of life is more of an entropy ('order') problem? One of the basic equations of chemistry is $\Delta G = \Delta H - T\Delta S$. ΔG is Gibbs 'free energy' and is suggested to determine catalysis, ΔH is very much the kinetic component, and $T\Delta S$ the entropy ('order') question and which is affected, and made more difficult, by higher absolute temperatures.

Certainly, water molecules in ice are fixed in position; in the liquid phase, they are increasingly mobile as the temperature increases, and as gas molecules, they can really fly about more or less independently. This is all in agreement with the high temperature dependence of the entropy component. So there really is an important problem—is life more of an entropy question or a kinetics question? It would be very good to know!

Thus thermodynamics might favor a lower temperature origin of life. Unless the origin of life is really a 'kinetic' problem. But in the opposite of a Popperian perspective, researchers appear to have quite fixed ideas for the temperature for the origin of life! It appears that many people are convinced of a higher temperature origin.

I (and almost everyone else) have only dealt with the question quite superficially, but I do favor a lower temperature origin—I guess that we consider life to be more of an entropy problem and not so much a kinetic problem. We have done some work on the question, and that was on the extent of RNA-folding versus temperature. For example, we could use tRNA, which normally are folded; and give a circular dichroism (CD) signal sequence, which detects secondary structure. And similarly, we could take any random sequence of RNA and see how well it folds (in the computer). The results of the latter experiment are in Figure 4.3, folding of RNA structures versus temperature. So yes, at the very highest of temperatures (say, above 80°C), we would not expect RNA molecules to fold properly (at least before proteins were invented). And there is a strong temperature effect from increasing the GC content (which has three hydrogen bonds, not just two). It does appear that the RNA-world scenario supports a lower or intermediate temperature.

At Cambridge University, they have studied reactions at freezing temperatures (–1.6°C) and have found that the reactions are carried out at this temperature (Attwater et al. 2013). This supports (but certainly does not prove) that the entropy question is important. So maybe life did originate at lower temperatures! In contrast, Lane et al. (2012) favor a high temperature origin of life, and this is probably the majority view—life is a 'kinetic' problem. These authors study the properties of membranes at high temperatures, and this should give some fundamental insights into membrane properties irrespective of the temperature at which life arose! So whatever the temperature life arose at, there will be some useful information from other scenarios. But dammit, we still do not know the temperatures at which life arose and whether it is more of a kinetic problem or an order (entropy) question. It would be good if we did know—some experiments could really help here.

It would be helpful if there was a cycle of reactions that catalyzed each other; for example, if A catalyzes the reaction B → C, C catalyzes D → E, D catalyzes reaction F → A; these are called autocatalytic cycles. Mathematicians Hordjick and Steel (2013) have shown that if there is randomness as to what reaction catalyzes another reaction, then we would expect that some autocatalytic cycles occur. So that is an optimistic part of the picture—autocatalytic cycles are to be expected.

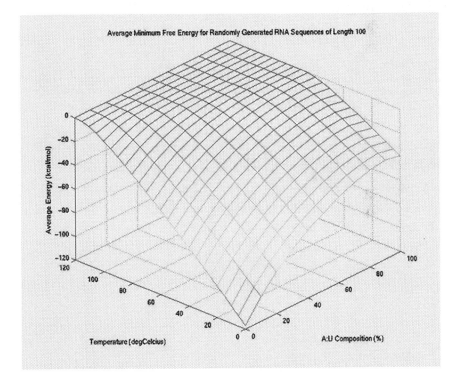

Fig. 4.3. RNA folding at different temperatures. As the temperature is lowered, the gain from folding increases (becomes more negative). There is improved folding at lower temperatures, especially with higher G:C composition (which has three hydrogen bonds, not two as in A:U base pairs) (from Moulton et al. 2000).

We will not discuss here the suggestion that life began somewhere else in the universe and then was later transferred to earth. This does not really 'get rid of the problem' of, for example, the temperature problem of the origin of life. There are many places in the universe that are relatively cold (below 0C), but I have not seen any statistical treatment of the average temperatures related to the possible origins of life (there must be some). It does seem likely that there are many more places at lower than at higher temperatures. Such ideas would be helpful—it would be a good problem to solve. However, if life did begin elsewhere in the universe, then presumably it only got to earth only once because all life has the same code.

Summary

Overall, we see that the twentieth and early twenty-first centuries have been very productive. There has been the development and the acceptance of genetics with its linear arrangement of protein coding sequence. But we still have a long way to go in determining the tertiary structure of proteins and especially their assembly into larger units (and cells). And there is still the question about the regulation of genes and the folding of RNA molecules. We have a lot to learn here, and that is good news.

CHAPTER 5

Human Evolution

Humans are very interesting, and it would be easy to (and people have) write whole books about the evolution of humans. But in this chapter, after a very brief introduction, we will consider only two aspects of human evolution. These are, first, whether there is any truth in Descartes's division into physical matter and mind matter (you will remember that this was a seventeenth-century distinction that did give scientists some areas of authority, though in strictly limited areas). And second, the African continent was the ultimate home of humankind.

But first, we give a very brief introduction to humans—on a genetic or genomic scale, there is nothing unusual about our genome. It is a 'normal' primate one. Figure 5.1 lists some of the differences between a chimpanzee and a human genome. But for example, the chromosome number of 46 is two less (a maternal and a paternal one) than many of our non-human relatives—so there is 'one chromosome fusion'. As the figure reports, these are all normal differences that occur even between populations. So in that sense, humans are quite normal. But we should get back to the two main questions of this chapter.

- ❖ one chromosome fusion ☑
- ❖ some enzymes lost or gained ☑
- ❖ some differences in copy number ☑
- ❖ many small inversions ☑
- ❖ transposable elements activated ☑
- ❖ many indels (insertions/deletions) ☑
- ❖ many point mutations ☑
- ❖ some introns expanded/contracted ☑

Fig. 5.1. Genetic differences between humans and chimpanzees. All differences between the species are the normal types of differences found within populations. Genetically, humans arise quite 'normal' on this data (indicated by the tick).

Descartes's Suggestion

So we must consider Descartes's question—as we have already mentioned, he suggested in the seventeenth century that there was 'mind' matter and 'physical' matter. Although many later authors did not accept this subdivision (including Charles Darwin in the nineteenth century), it does seem that many people accepted this hypothesis well into the twentieth century. What would we have to demonstrate in order to suggest that there was some fundamental difference between the human and animal minds that could not be breached by normal mutations? Human languages are a good candidate. Is there any evidence that the great ape and the human minds are qualitatively different? Let's see.

There were some early experiments in the Canary Islands on the ability of chimpanzees to solve quite complex problems at the Tenerife Prussian Academy of Science from 1912–1920 (see Köhler 1973). Fortunately (perhaps), some people from Germany were stuck on the Canary Islands during the First World War and had to continue their experiments! Most of these experiments were on showing that chimpanzees had mental abilities well in excess of dogs, cats, and rats; and most used the rewards of fruit. For example, the chimps could use quite elaborate structures, including using boxes, climbing structures, and using sticks to get bananas or pears from difficult locations. However, some early experiments in the Canary Islands did consider language ability; and from now on, we will just concentrate on the language ability of chimpanzees (both species) and gorillas, genetically the closest relatives of humans.

Early attempts to teach chimps to speak ended with them only able to utter a few sounds (for example, with the chimp Viki). It was quickly realized that they did not have the physical characteristics (voice box and vocal cords) for human-like speech; therefore the experiment said nothing about the potential mental abilities of the chimps. For humans we consider people who cannot speak as 'handicapped' and does not affect their status as human beings. Furthermore, an important point here is that chimps do make over thirty 'sounds' to communicate in nature, and we could consider these as words. But more importantly, it was then found that they can be taught sign language and become quite proficient at this nonverbal language. Thus we need to separate 'language' ability from 'speech' ability. Consequently, later attempts at teaching 'language' have concentrated on other representations of language, such as initially with ASL (American sign language) and then later plastic tokens on a magnetized board, computer keyboards attached to a voice synthesizer, writing, etc.

How do you decide the most effective way to try teaching language to chimpanzees? The most success is in following the model of language learning in young children. The Gardner's raised a baby chimpanzee (Washoe) in their home, and most subsequent studies have followed this pattern to the extent (as with children) of starting as young as possible and having considerable everyday contact with humans. Others have had similar success with gorillas. Less work has been done with orangutans, but after a fast start (and even exceeding some humans), they may stop development earlier (and end up less intelligent than chimpanzees and gorillas, see brain size later). Much of the early difficulties was 'human' in the sense that we had not learned the most effective ways of communicating and testing the great apes. As these difficulties are reduced, we have been continually surprised at the mental abilities of the great apes.

However, the 'Clever Hans syndrome' was one experimental difficulty. This is named after a horse that responded to inadvertent cues given by its trainer. For example, if asked 'What is 6 × 4?' Hans would tap one foot repeatedly until the trainer, by almost imperceptible body language, indicated the correct number of taps had been made. The trainer was not aware of these signals and believed the horse could indeed count. It is now routine to test for the 'Clever Hans' effect with, for example, the questioner not knowing the answer to a question being posed and an independent observer watching through a one-way mirror.

It was quickly learned that the great apes (chimps, gorillas, and orangutans) could learn sign language. Most of the training has been done with ASL (American Sign Language), but the main thing is that they can learn it. So basically, language is not a problem as long as we start them young, just like children! Over 1,000 signs may be learned by chimps (genetically the closest to humans).

So to what extent then do chimpanzees and gorillas have the ability to learn 'language'? Let's examine some components:

1. Representation by 'abstract' concept, symbol or sign (signs are arbitrary in the sense that they differ between countries and languages), plastic tokens, or computer keys. As we have said, over 1,000 signs might be learned.
2. Can name general concepts (for example, 'birds').
3. Can form new combinations of signs for a new experience (for example, 'water bird' for a swan, 'baby truck' for a large station wagon/high-sided trailer combination).

4. Can ask questions and can use 'who', 'what', 'why', 'when'.

5. Have the ability to decode spoken sentences. This is a major achievement from the viewpoint of artificial intelligence for computing systems; it has been a major activity to get computers to decode speech accurately. Spoken sounds are not formed in a discrete manner with pure consonants followed by pure vowels— rather, they overlap, and we decode them into syllables. Apes have this ability, presumably from their normal listening ability, which is routinely analyzing sounds in the forest, including the over thirty sounds they use in nature.

6. Can form words into short sentences. There is some problem here in understanding how complex the word order can be in ape sentences in ASL. People like Noam Chomsky (more of a cynic perhaps) placed great stress on grammar, but word order is a more flexible in ASL (and sign languages generally).

7. Can tell lies. For example, Koko (a gorilla), unaware she was being observed, accidentally broke a toy. Later when asked who broke the toy, Koko blamed the other younger gorilla. (Younger children in a family might recognize this behavior from their older siblings, though older siblings usually deny that they would 'ever have done that'!)

8. Have self-awareness in the sense they can recognize themselves in mirrors and in photographs. For example, in an experiment, a white mark was made on a chimpanzee's forehead while it was under anesthetic. On seeing the mark in a mirror, the chimpanzee immediately rubbed its forehead to remove it.

9. Have considerable ability in solving puzzles, such as stacking boxes on top of each other or inserting a narrow stick inside another in order to reach a banana.

10. A more advanced criterion is 'the theory of mind'. A great ape works out that other great apes or humans are like their own and adjust their behavior accordingly. In one experiment, a chimp was let out early from its overnight quarters, shown where a

banana was hidden, then taken back to its accommodation. When all the chimps were let out, the first one encouraged the others to feed in a different part of the enclosure. After they had all fed and the others were resting, the first chimp snuck off, got the banana, and ate it all by itself!

Is all this latter ability language? Probably irrelevant! We can 'define' language in many different ways—either to include or to exclude these abilities of the great apes, but we must not exclude humans who, in some way, are 'handicapped'. The important thing is to examine the components that go into human language, then see which of these components are recognizable in the great apes. Remember the quote from John Maynard Smith about the use of words or questions about the world. It is useless to get involved in discussions about whether apes have 'language'—that is what 'philosophers' might do? Would they? Surely not! What is important is to be able to analyze and measure the components that make up language—that is the way scientists should do it.

The following is an early example of experiments and is from the training of a young female gorilla, Koko (Patterson and Linden 1981). She also began to answer questions posed. One day a visitor asked what the sign for 'good' was. Before Patterson could respond, she noticed Koko making the sign 'good'. So Koko had understood the spoken sentence and made the appropriate sign. The trainers started to use the device of spelling rather than saying key words when they didn't want Koko to know what they were talking about. However, Koko still worked out that C-A-N-D-Y spelled a favorite treat, so they had to use other subterfuges when discussing such topics. Koko's growing understanding of English had its practical uses as well. If Patterson's hands were otherwise occupied, she could always tell Koko to clean up her room and have her respond correctly by fetching a sponge and wiping up the mess she had made.

In a related study, they also tested the same young gorilla on IQ tests for children (that did not require a vocal response). Koko (and also young chimps) do quite well on this test—and they improve as they get older. Two questions of non-verbal IQ tests that are used with young children are shown in Figure 5.2. They might have an IQ of at least 85 when young, though some orangutans develop quite quickly and may even exceed young children for a short while. However, the apes do not show the continued development that young children do—we end up (fortunately) much more intelligent than the apes.

Which of the four lower selections best completes the series on the top?

Which of the lower boxes best completes the series on the top?

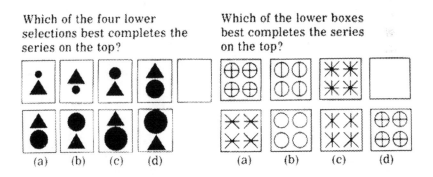

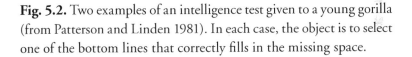

Fig. 5.2. Two examples of an intelligence test given to a young gorilla (from Patterson and Linden 1981). In each case, the object is to select one of the bottom lines that correctly fills in the missing space.

A related question is the continued growth in brain size of children (compared with other primates), and this is indicated in the figure following (Figure 5.3). This figure really explains many of the differences; humans just keep on growing their brains for much longer. So (fortunately) relative and ultimate brain size does seem important; it does seem that there are not any fundamental difference between ourselves and the great apes that would require 'special' mutations.

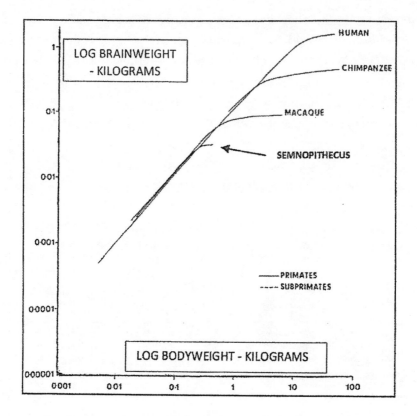

Fig. 5.3. Brain-body growth curves for four primates showing similar extended infant curves at early stages of growth. The monkey (*Semnopithecus*) was mainly measured in embryos and the chimps after birth. (Several other mammals (subprimates) are at a slightly lower trajectory.)

The great apes generally performed worse than children when a verbal response rather than a pointed response was required. When tasks involved detailed drawings (such as penciling a path through a maze) or precise coordination (such as fitting puzzle pieces together), an ape's performance was also inferior to that of children. So what have we leaned? Apes and children learn a lot while they are young, but young human children keep on developing. Perhaps.

Here we will only allow ourselves one more question about humans, and that is the related question of 'Where does the human skull shape come from?' This is shown in Figure 5.4, and it fits in with the idea that humans show accelerated development. But then growth stops, and we do not show the later stages of chimpanzee skull growth, with the large protruding jaw and the large brow ridges.

Fig. 5.4. The pictures are profiles of an infant and an adult chimpanzee. We must be a bit careful here; the infant chimp is not a living specimen, but a stuffed dead specimen (see text). Note the position of the ears, protruding lower jaw, and the eye ridges. The point is that the human skull shape was there all the time—but it was not 'lost' as we grew up!

We have to be a bit careful here; the photograph above of the young chimpanzee is not a living specimen, and it was 'posed' to emphasize its humanness. However, I was once visiting Wellington Zoo with some friends, including their 2-year-old daughter (and they had several young chimps there). So I had a direct comparison, and my friends had to agree that their 2-year-old daughter (of whom they were very proud) looked very much more baby chimp-like than adult chimp-like. So I was convinced about how the human profile came about.

So what have we learned about humans and about Descartes's division into 'physical matter' and 'mind matter'? The evidence is now overwhelming that normal mutations can lead from a chimpanzee-like ancestor to humans—there appears to be nothing about humans that normal mutations could not lead to a voice box and to the trajectory that leads to a larger brain size. So now it is time to consider our second question.

Africa as the Original Home of Humans

When I was a graduate student in the United States, the late Louis Leakey gave a seminar (many years ago now) on his discovery of early human fossils in Africa. At a reception held after the seminar, several paleontologists said, 'We know he is looking in the wrong place. *Everyone knows* that humans evolved in Asia.' Everyone knows? The molecular biologists were far more sympathetic. They said, 'Morris Goodman is showing that humans, chimps, and gorillas are all closely related—and because chimps and gorillas are African, we should be looking in Africa for early human fossils. Louis Leakey is looking in Africa, and he is finding early fossils.' So early on there were doubters and acceptors about humans arising within Africa. In the previous century, Charles Darwin (following the conclusions of Thomas Huxley, a famous nineteenth century anatomist) also concluded that humans had evolved in Africa. Based on anatomy, Huxley had found that anatomically chimpanzees and gorillas were morphologically the closest relatives of humans, and because these other two species were African, then the simplest explanation was that humans arose there also. Now it is common knowledge that humans evolved in Africa, and the brilliant Leakey fossils are all part of the evidence. The molecular, morphological, and now the new paleontological evidence all agreed; and Louis Leakey's fossils were quickly accepted.

Yes, it is the simplest explanation, but we could all think of other more complicated explanations; but until we get more evidence, we use the simplest, though (as we have said) we should never 'believe' any hypothesis uncritically. However, and as we have said, an African origin for humans was not generally accepted in the first half of the twentieth century—possibly for nonscientific reasons, such as wanting Europe (or as a second choice, Asia) to be the original homeland of all humans. Bowler (1986) reports on some early aspects of these early evolutionary ideas, which include quite a range (including that humans were extremely diverse, having independently evolved in several different locations). Well really, there was no additional evidence in the first half of the twentieth century. The more specific the prediction, the more useful the hypothesis, the easier it is to test. Now we see a 'consilience of induction' with similar conclusions coming from both paleontology, morphology, and molecular biology—good science. Thus it was soon accepted that Africa was the ultimate homeland of humans. But it was also found that later human fossils (*Homo*, but not *H. sapiens*) from between 1.5 and 2 million years ago were also found in both Europe and Asia. So we can have at least two models, although both groups now accept that the original homeland of humans was Africa, say, 5–7 million years ago. The two models are that modern humans evolved within Africa and moved out only 50,000 years ago, and the other is that they moved out of Africa about 2 million years ago. Let's look at the evidence.

The Recent 'Out of Africa' Hypothesis

Research of New Zealander Allan Wilson's lab in Berkeley, California, on comparative human genetics started quite early. They originally used the quantitative microcomplement fixation method to test that there was no apparent slowing down of human albumin evolution since the time of the divergence of apes and humans. Hence they suggested that the great apes and humans have a more recent common ancestry (about 5-6 million years ago) than the 20–30 million years ago that was originally supposed. This early work was measuring the strength of the immunological cross-reaction between the same functional protein

from different species. In this early paper, they could not tell the order
of divergences of chimps, gorillas, and humans and left it as a three-
way split (a trichotomy). Later work showed gorillas separated first, and
then chimps and humans split (maybe a million or two years later).
Chimps later split into two—the common chimpanzee and the bonobo.
The main point here is that humans are not a very ancient lineage
from 20–30 million years ago. However, the African origins and 5–7
million years ago origin of humans is accepted by both the recent 'Out
of Africa' and multiregional models. We should also say that (although
we will discuss mostly molecular data here) most paleontologists also
supported the recent African origin—for example, Chris Stringer of the
Natural History Museum in London. He long ago favored the origin
of modern humans in Africa and saw a yo-yo effect in the Near East
as the climate improved, with an effect in the Near East of advances
and contractions of the early *Homo sapiens* expansion out of Africa (see
Stringer and Andrews 1988).

So the 'recent "Out of Africa"' hypothesis is that modern humans
evolved in Africa and that a subset of them moved 'out of Africa' on
more than one occasion, into Asia and Europe and eventually to the
rest of the world. On such a model, genetic diversity is predicted to
be highest in Africa and lesser elsewhere, which is what is found. The
model is often considered as dating from 1992 (Wilson and Cann
1992), though some of the information was available earlier. Their
model also envisages that earlier forms of *Homo* that migrated into
Asia and Europe did not evolve into modern humans even though
(as closely related 'species') there could have been some interbreeding
(that geneticists call introgression, and this is observed in the case of
Neanderthals). There is also another group called Denisovans (who
appear to be related to Neanderthals), and they have left a small genetic
marker effect on people in (at least) Melanesia.

In other words, basically, these very earliest fossils from Asia and
Europe have no direct descendants. However, do be careful here; the
phrase 'Out of Africa' can in principle also be used to describe the

very earliest *Homo* movement from Africa into Asia and Europe, the later Neanderthals (and Denisovans), or even later modern humans. When I refer to the 'recent "Out of Africa"' model, as contrasted to the 'multiregional model', I am referring to the origin of modern humans (and still allowing some introgression from groups such as the Neanderthals and Denisovans), but not to the earliest group of *Homo* that came out of Africa).

Multiregional Model

The other end of the spectrum of ideas is the multiregional model (Thorne and Wolpoff 1992), which still accepts that humans originally evolved in Africa but suggests that these early humans (in Africa, Europe, and Asia) have lived in these regions for well over a million years. So if humans were a very old population that had existed for more than a million years in each of Europe, Asia, and Africa, then basic population genetics tells us there would be relatively high genetic diversity, both within and between the populations.

On this model, you would expect that there would be some very old genetic diversity in each of the continents. You should think about testing these other models along the spectrum of ideas, though we will still concentrate on these two main models. So what I will do here is to look at how the research has given us new information, particularly in relation to the two ideas along the spectrum. Good science must always avoid just accepting models without testing them.

We now have a start, and geologically speaking, humans are not an extremely ancient lineage (though 5–7 million years ago is still pretty old, but certainly it is not 25–30 million years ago of separation). However, the Wilson lab in 1967 used a fossil calibration point of about 12 million years ago for the divergence of the *Sivapithecus*, an Indo-Pakistan fossil probably on the orangutan lineage, but that time might be too recent. We have suggested that the lineage got into Africa more like 17–19 million years ago, when the Suez region was more amenable to the dispersal of apes into Africa—this is why we say 5–7 million years ago, and not 5 million years ago. But both the recent 'Out of Africa' and the multiregional models accept an African origin for the genus *Homo*. The next paper we discuss from the Wilson lab is over a decade later, and it reported that humans had low genetic diversity (Ferris et al. 1981). Ape species are two to ten times more variable than the human species with respect to the nucleotide sequence of mitochondrial DNA even though ape populations have been smaller than the human population for at least 10,000 years (genetic diversity does increase quite slowly). The amount of intraspecific sequence divergence was greatest between orangutans of Borneo and Sumatra. The least amount of sequence variation (for apes) occurred among lowland gorillas, which exhibit only twice as much sequence variation as in humans.

But the main point, and in contrast to the multiregional proposal, was that humans were all genetically very similar. That relatively low genetic diversity among humans is pretty striking! The lineage to modern humans does not seem to have been in Africa, Asia, and Europe for over a million years—or else they would have very high genetic diversity. At that time, the explanation was based more on a reduction in size of ape populations; these days we would focus on the rapid increase in human numbers over the last 100,000 years or so. So we can add this support from the low genetic diversity to the 'Out of Africa' model and get the following: Out of Africa, 1; multiregional hypothesis, 0.

The formal description of the multiregional model is often taken as the Scientific American article of Thorne and Wolpoff (1992). But you will see that the Ferris paper was a decade earlier; it was already known that humans had a low genetic diversity. So in a sense, this point to Out of Africa is almost an 'own goal' for the multiregional model. But we had better be careful here; these results contradict the multiregional model of a large population size over a million years but do not yet focus specifically on Africa. If this was all the information we had, we could not distinguish a founding human population in Asia, Africa, or Europe. Humans only got to Australia about 50,000 years ago and to the Americas a bit over 14,000 years ago, so we can leave out both those continents (and Oceania) for now. The next piece of evidence is also from Allan Wilson's lab (Cann et al. 1987). Cann used the newly available techniques to measure directly some of the genetic diversity in mitochondrial DNA within and between human populations from around the world. Her results did include many people from Africa, Asia, and Europe, so she could focus on each continent in turn. But be very careful here—talking of 'Eve' does *not* imply there was only one woman in the community! Indeed, other estimates are around 5,000–10,000 women (and presumably a similar number of males). But just by chance and given a long time, only one version will persist. For example, a woman may have three sons reaching adulthood, but her mitochondria will not have been passed on. (The length of mitochondrial DNA is only about one part in 100,000 of your total nuclear DNA, but it is maternally inherited—both sons and daughters get it from their mothers). Anyway: Out of Africa, 2; multiregional, 0.

The work of Becky Cann used enzymes that recognized just short sequences of DNA. The next test that came after DNA sequencing became routinely available, and even though it was initially difficult (and only a few hundred bases/nucleotides could be sequenced), it was a more direct approach, allowing a more quantitative testing of hypotheses. So DNA sequences from many humans were then published from Allan Wilson's lab, namely Vigilant et al. (1991). This showed even more strongly the higher genetic diversity in Africa and lower diversity outside Africa. So this was another test of the hypotheses, and so we now have Out of Africa, 3; multiregional hypothesis, 0.

So the 'Out of Africa' model has passed three tests, and although everyone has mitochondrial DNA in their cells, it is only inherited from mothers. So perhaps we should be careful and say all the mothers came out of Africa, but what about the fathers? Given a possible slower rate of geographic dispersal in the distant past, is it not possible that the males would be different? But let's check the Y chromosome, which is only found in males. Although it is unlikely, maybe the males could have come from somewhere else (no, not Mars). This took somewhat longer because nuclear DNA is slower evolving than mitochondrial DNA, and techniques had to improve in order to get longer DNA sequences. However, when Ke et al. (2001) studied 12,000 Y chromosomes from around the world, they found that they also had their highest diversity in Africa. So the 'Out of Africa' model has passed another test: Out of Africa, 4; multiregional hypothesis, 0.

So we now have all the mothers and all the fathers coming 'out of Africa'—isn't that enough? But some people just can't stop testing ideas even more thoroughly. What about all the other autosomal genes, those in the nucleus, but not on the sex chromosomes—that is, not on the X and Y chromosomes? That required even more improvements in techniques, but they have now been sequenced and analyzed (Tishkoff et al. 2009). Again, we find the highest diversity within Africa (and remembering that human genetic diversity is relatively low), so Out of Africa, 5; multiregional hypothesis, 0.

But you still can't stop some people—good. Another test was to sequence the whole mitochondrial genome (c.17,000 nucleotides) from Neanderthals. On the multiregional model, this should be within the diversity of modern humans (because of the suggestion that Neanderthals gradually evolved into modern humans, perhaps with some admixture of genes from other populations). But no, the Neanderthal mitochondrial genomes were well outside the diversity of modern humans, including those of modern Europeans. The reference given here is to a group of five Neanderthal mitochondrial genomes (Briggs et al. 2009) because that allows a comparison with the full diversity of Neanderthals, giving an even stronger conclusion. So now we have Out of Africa, 6; multiregional hypothesis, 0.

Surely that is enough, but the latest advance is from 2010—the publication of the full Neanderthal genome (Green et al. 2010). Again, that shows that most nuclear DNA of Neanderthals is outside that of modern humans: Out of Africa, 7; multiregional hypothesis, 0. Oops, we need to be careful here; there is a small amount of introgression from Neanderthals (though not from earlier *Homo* groups). Let's follow this up.

It is useful to note here that much work on Neanderthal genomes is done in Leipzig (Germany) in the lab of Svante Pääbo. He developed the initial techniques for ancient DNA while working as a postdoctoral fellow in Berkeley in Allan Wilson's lab. So there is another link there, and science is truly international. As an aside perhaps, the Leipzig group always seemed keen on finding evidence for interbreeding between Neanderthals and early modern Europeans. Geneticists call this 'introgression', where by interbreeding some genes pass between related 'subspecies'. There is about a 2%–3% introgression of Neanderthal genes into modern Europeans (and eastern Asians), and possibly it is related to diseases endemic to the region. However, there is nothing in any introgression that makes us distinctly human—the fossil evidence is that modern humans are identifiable in Africa 50,000–100,000 years before they arrive in Asia and Europe. Maybe we should allow the multiregional model half a point (or perhaps 0.03 of a point)! Oops, perhaps we should also say here that there is a temptation to overclassify early *Homo* species, people seem to want to 'classify' a new *Homo* species!

The recent 'Out of Africa' hypothesis for origin of modern humans makes testable predictions, and these predictions keep passing new and more specific tests. In contrast, the multiregional model keeps failing test after test. So should we consider the 'Out of Africa' the Gryffindor model and the multiregional model the Slytherin one? But no, we should never be too judgmental. You can hear real scientists say, 'XXX was a good hypothesis. Yes, we have proved it wrong, but it led to us devising several new tests in order to do so.' That is a much more positive approach—never 'believe' your models, but use them to devise more and more specific tests. So remember 'belief is the curse of the thinking class'. Okay, we asked our first-year student to identify both sources of this 'quote'. Karl Marx said 'drink is the curse of the working class', and Oscar Wilde reversed it: 'work is the curse of the drinking class (by which he referred to the aristocracy, to whom paid employment was an abomination—at least according to Oscar Wilde). But the main point is never to believe your models, but to test them.

Yes, there is some (recent, a few thousand years ago) back flow of Central Asians/Europeans into northern Africa, but that is comparatively recent. So, what have we learned about humans? First, at the moment we cannot find anything that violates the 'continuity of mind' between a great ape ancestor and modern humans. Second, we have come back to Charles Darwin's understanding (through Thomas Huxley) that humans first arose in Africa, and that modern humans have arisen there. Yes, modern humans have come 'out of Africa' several times, and there is some introgression from Neanderthals into Europeans and central Asians (perhaps we tend to 'overclassify' humans). Overall, we now have a very good idea about human evolution (where it occurred and when), and very importantly, we can also reject Descartes's division of physical matter and human mind matter. This was a useful distinction in its time in that it allowed scientists authority in a limited area. But it has long ago served its purpose, and we no longer can make the distinction. Of course, we mustn't 'believe' our hypotheses, but we would now be surprised if anybody found some feature that could not be explained by normal evolutionary processes.

There is one lesson I will comment on here, and that is that we need both specialization and generalization in science. Yes, we need specialists in different areas, but we always need to be aware of good data from outside our area of specialization. There certainly is a moral here: good communication between different areas of science is extremely important even if we work in our own specialty most of the time.

CHAPTER 6

One Idea that is Probably Wrong: The K-Pg Extinction of Dinosaurs

If scientists really are human, then it is to be expected from a Popperian viewpoint that they (the scientists) will be 'believing' some erroneous ideas. But which ones? DNA only knows (and s/he is not telling). The problem discussed here ultimately relates to the impact that of the asteroid that hit the earth 66 million years ago, and that marks the end of the Cretaceous and the beginning of the Paleogene (the K–Pg boundary, historically the Cretaceous–Tertiary boundary, K–T). Did this impact really kill off the remaining dinosaurs, or was it normal replacement (of the expanding mammals (and birds)) that did in the dinosaurs? We first consider the recent extinction of megafauna (over the last 50,000 years) and find that this is attributed to an invasive species—humans. Why the big difference—an asteroid for the fifth mass extinction, humans for the sixth? This is the question we investigate.

In this chapter, we will do three main things: First, we will discuss some general questions about mass extinctions. Second, we will consider the most recent 'mass extinction', starting about 50,000 years ago when modern humans arrived in Australia and then later in other parts of the world. Third, we will get to the real crux and finally examine the K–Pg extinction of dinosaurs—what really happened? The hypothesis that modern humans caused the elimination of the existing megafauna is excellent. Is this the first time a new evolving group caused the elimination of an earlier group? We will see that very different mechanisms are proposed for the sixth and the fifth mass extinctions. Why?

What Do We Expect of 'Mass Extinctions'?

There are several reasons why the arrival of placental mammals (in particular) are probably significant factors: relatively larger brain size (with more flexible behavior), teeth for chewing their food, living off smaller reptiles, parental care of their offspring, etc. Is there any progress during evolution? Or are the oldest animals able to fully compete with all the later animals? (Or earlier plants with later plants for that matter!) There are still other possibilities, but the main thing is that we must test our hypotheses and not merely 'believe' them (and then decline to test them). There is a strong 'belief' that an asteroid impact killed off the dinosaurs, but this is probably wrong, though we must always be willing to test our ideas.

The question is crucial because it reflects on the processes that are important in the science of evolution—and the impact hypothesis assumes some currently discredited ideas from early studies of evolution, in particular impacts (catastrophes) 'driving' evolution. (Ouch, we do not like the term 'driver'; there is no good evidence that evolution needs external 'drivers', but that really is a separate issue.) For example, we have seen that Cuvier (who, on the positive side, worked out much of geological stratigraphy) but thought that animals were fixed in form over a geological period, and then there were 'mass extinctions' by external catastrophes, and that this reset the animals available for fossilization. This is quite an important example because Cuvier was correct about some things and wrong about others—he was human! But catastrophes?

We mainly concentrate on the extinction of land-based groups, partly because this is also the focus of the loss of the megafauna during the early phases of the sixth mass extinction, and unfortunately, it may have included Neanderthals as a distinct group (even though there was some interbreeding between Neanderthals and modern Asians and Europeans). It also appears that most of the losses at the end of the Cretaceous were also among land-based organisms. However, the early paper advocating an extraterrestrial impact as a cause of the mass extinctions (Alvarez et al. 1980) also suggested a major effect on both pterosaurs and several groups of marine reptiles, so we cannot neglect them entirely.

We will first examine the relatively recent loss of the megafauna over the last 50,000 years because it is now generally accepted that the arrival of humans eventually led to the extinction of this megafauna. The first point is that humans had arrived before the extinctions (see Figure 6.2 later). Additionally, there were many Ice Ages extending back ~2.6 million years; there are nearly forty Ice Ages leading to nearly eighty major changes in temperature—but without 'mass extinctions until the last one. There was *cooling* of the environment in Australia during the onset of the last Ice Age, but *warming* of the environment after the Ice Age for North America and for the Polynesian expansion. These Ice Ages are very important as controls and should really eliminate any concept of climate change being the primary factor even though climate changes must always, always be important. And we have already seen that we cannot 'blame' early humans for these extinctions—they probably assumed that there was continued spontaneous generation and that the animals they killed would be regenerated. Okay, we now realize that 'extinction is forever' (well, unless we can regenerate them), but we have really only learned this over the last 200–300 years. However, we really don't know about early human 'beliefs', particularly those from 10,000–50,000 years ago.

Currently there is also interest in the very recent geological period, the Anthropocene, when humans leave identifiable signals in the geological record. The signal varies considerably, but are very real. The signal includes the earlier disappearance of large fossils, an increase in CO_2 concentration, and an increase (hopefully a simple spike) in radioactivity. A natural question is whether this is the first time a new species or group of species have led to the recognition of a new geological period, or has it happened previously when new groups have arisen and eventually replaced earlier groups? We still need to test other factors, such as whether normal microevolutionary processes allowed the relatively larger brained mammals and birds to replace the dinosaurs. There can be many factors involved, and it is unusual that just one (the impact) has been singled out for the demise of the dinosaurs.

Before starting on the most recent (sixth) mass extinction, we need some background information about extinctions. For the first five, there is perhaps general acceptance that they were primarily initiated by physical factors, although this was not always the case. Mass extinctions are defined as exceeding some 'background', though usually without analyzing the effects of newer groups of animals that may alter the competitive landscape. It was usually assumed that normal microevolutionary processes could not account for these extinctions above the background rate and that therefore they must have different factors 'driving' them. Though this 'explanation' clearly does not follow without testing, it could be the arrival of new groups of taxa. This assumes differential affects at different places as the new groups arrive. Perhaps we should expect (predict) that the expansion of new groups of animals, such as humans or placental mammals, would have major effects on earlier animals. Is there any progress in evolution? We need to be careful here—evolutionists do not accept that evolution is 'progressive' (at least not in the shorter term), but we do need also to consider longer-term effects, in particular the origins of new groups of plants and animals. We must have paleontological, morphological, molecular, behavioral, and ecological data integrated into one testable picture of evolutionary change and allow eventual progress in evolution.

This aspect of primarily considering non-microevolutionary factors was not formally tested for the first five mass extinctions, and the option that increases in extinction rates could be explained by, for example, new 'invasive' groups of taxa remains. An interesting contribution has been made by Lynch and Abegg (2010, see later) who point out that complex adaptations (requiring more than one mutation that are neutral of themselves but have a positive effect on fitness when combined) are very much more likely to occur in larger populations than in the smaller populations (that are characteristic of very large animals or of trees). This might explain the observed long-term turnover of the largest taxa (including trees) and is discussed again later under Cope's rule. This is supplemented by the results of Zhong et al. (2014), who report that very long generation times only appear compatible with lower mutation rates. So new developments might occur in the larger populations of smaller- or medium-sized plants and animals with higher mutation rates. This needs to be examined.

There is also some evidence placing the last common ancestor of placentals in Gondwana (see Romiguier et al. 2013), where there is a much poorer fossil record at that time—no mammal fossil has apparently been found in Africa in the late Cretaceous period between about 100–66 million years ago. It is important also to resolve the different estimates between fossil and molecular data, including the quality of the fossil records in different locations. There are three important points here. The first is that the quality of the fossil record is always important; there are regions and times where the fossil record is excellent, but there are other regions and times where it is incomplete or missing. For example, in Australia the three main Paleogene and Neogene (post-Cretaceous) sites are Murgon (about 55 million years ago), Riversleigh (about 12–25 million years ago), and Etadunna formation (about 26–20 million years ago). So outside those times, there may be little good information. We must consider all our sources of information, including the lack of good fossils from particular areas at particular times (e.g. early primates were apparently missing from South America until recently). It is definitely not adequate to draw worldwide conclusions based primarily on the records, for example, of just North America (and possibly China). This is more general than the important consideration of the rarity of taxa (the Signor-Lipps effect); this is about worldwide distributions rather than local.

The second point is what the current crown group (Figure 6.1) of either mammals or placentals or birds has to do with events happening at the K–Pg boundary 66 million years ago is quite unknown. For example, multituberculates would have been in that crown group of mammals 66 million years ago at the time of the K–Pg impact and probably until about 35 million years ago. It is certainly accepted that the crown group changes with time.

The third point is that, as we tell our undergraduates, morphological and physiological data are too important for just finding the phylogeny—it should help tell us about adaptations that have been occurring. There are many times during evolution that we have had morphological or physiological convergence, and it is important to find these cases and to understand their principles. We expect to find the underlying tree from (the more neutral) molecular data and then to use morphological data to find the examples of convergence. But we do need to know more of the reasons why molecular and fossil data can give different estimates (and possibly not just a crown group/stem group difference). Ultimately, we must be able to integrate all our data—fossil-based, molecular-based, genetics-based, and ecology-based.

We address here two important questions that cover two ends of a spectrum of ideas, and they are whether the mass extinctions

1. are abrupt ('sudden and unexpected') or
2. occurred over an extended period of time (which could vary from several thousand to many millions of years).

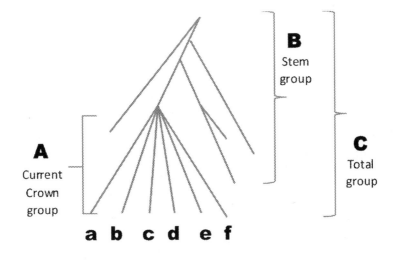

Fig. 6.1. The concepts of crown (A), stem (B), and total (C) groups, though illustrated for eukaryotes. The current 'crown' group is to the current last common ancestor and all of its living descendants. There may be up to six groups of eukaryotes, 'a' to 'f'. But the branching order (and the number of them) is currently unknown, though there are suggestions! The 'total' group is the union of both 'crown' and 'stem' groups. All groups vary with time.

Indeed, it appears that the latter (extinctions occurred over an extended period of time) is a significant opinion among paleozoologists for the fifth mass extinction (see the list of twenty-nine authors for Archibald et al. 2010). Sullivan (1987) goes as far as to suggest that the fossil record is 'unequivocal' in concluding a more gradual extinction. Indeed, Sakamoto et al. (2016) report that dinosaurs were in decline for tens of millions of years before the boundary (and see later). Some possible support for this interpretation comes from Wilf and Johnson (2004), who report some changes in plant fossils before the boundary. For the sixth (the current and ongoing) mass extinction, the majority scientific opinion favors the second option of the continuum—extinctions over an extended period of time, including into the future. For extinction to continue over an extended period makes the final extinctions to some extent predictable (and/or avoidable), and this is a contrast to the first example (the fifth), where many think of the extinctions as being 'abrupt' or 'sudden and unexpected'.

A likely method of interaction is from Pond (1977). who points out that both birds and mammals are distinctive in that they feed their young until the young are able to feed in the same niche as the adults. This allows an increase in optimization for a single niche. In contrast, many earlier large land vertebrates may occupy different niches at different stages of their development, thus inhibiting effective optimization for just one niche. We do expect intermediate states, for example, in parents looking after their offspring. However, it does also appear that there is some progress during evolution.

Certainly, there are important aspects of the physiology of large size shown by the larger dinosaurs, but perhaps this only emphasizes the different phases that a large dinosaur may have gone through at different stages of its growth. Codron et al. (2012) discuss some aspects of the huge differences in size between hatchling and full-grown adult dinosaurs and relate this to eggs being limited to a relatively small size (by diffusion requirements). A useful current example might be the Komodo dragon that occupies at least three phases during its life (Murphy et al. 2002), starting by the young living in trees and only later coming down to the ground. At an intermediate stage, it has to be careful that it does not get killed and eaten by larger Komodo dragons! It was probably an unintended consequence that the promising start, including some biological factors, was effectively lost after the publication of the Alvarez et al. paper (1980). These authors did not refer to papers that considered the biology of rates of extinction nor to the rise of birds and mammals. The paper (understandably) only gave the evidence for the impact—that was its aim and a point we agree with completely.

However, as part of understanding the potential biological interactions, we need to study modern-day interactions between, say, mammals and reptiles in order to understand the extent to which Charles Lyell's original dictum (for *Geology*) that 'former changes' are 'referable to causes now in operation', which included all causes that could be studied in the present even if they might have acted sporadically (such as impacts) in the past. However, it is interesting that although Lyell fully accepted the need for mechanistic approaches to explaining the past, he did not initially accept the idea of any 'progressive' aspect to change through time (evolution)— perhaps until later when Darwin convinced him. So can we learn about the biological and/or physical processes from modern-day experiments and observations that could help explain past events?

An important point is that there are many examples where there have been major physical events that appear to have had little long-term effects on the biota—in other words, there were no mass extinctions. These are 'controls' on what might happen (or not happen) as a result of a major physical change. One simple example is the 'resilience' in Europe of both Neanderthals and modern humans to a major explosive volcanic event. Similarly, a major eruption in Malawi about 75,000 years ago appears to have had little effect on early human diversity. And a much older example (around 135 million years ago) is a major period of volcanism that again apparently had little or no effect on 'mass extinctions'. In both these examples, the major events were volcanic, but a more general example is one view of sea level changes over an extended period of 100 million years. There are many major changes in sea level where there are no 'mass extinctions'. In relation to extraterrestrial impacts, Onoue et al. (2012) report examples from the Triassic period, when there are impacts apparently not related to some mass extinctions. We come to another example later where the temperature changes during the onset of Ice Ages do not automatically lead to mass extinctions of megafauna. This is a very important point—it is not sufficient just to find a major change in a factor and then just to assume that there is a major effect on the biota. We do need control studies.

These studies show 'beyond reasonable doubt' that changes in climate, though important in themselves, do not *guarantee* a mass extinction. It is certainly not sufficient just to assume there are biological consequences of every major physical event. We must also consider the timing of the biological changes to test whether there is indeed any 'mass extinction' as the result of any particular major event. Basically, there is no guarantee that a catastrophic physical effect must inevitably have major long-term biological consequences, as opposed to new biological developments. But of course, particular cases remain a good possibility for testing.

The Sixth Mass Extinction

The sixth mass extinction is the ongoing one, starting with the loss of megafauna at many sites around the world after (and only after) modern humans arrived for the first time (see Figure 6.2). As mentioned earlier, it is often considered to start after humans arrived in Australia, though recent work indicates an earlier effect on African carnivores from around 2 million years ago. Relatively early, Martin proposed the 'overkill' hypothesis (see, for example, Mosimann and Martin 1975), and there may be times when hunting by itself is sufficient to account for the disappearance of some of the megafauna, especially when, as is usual, the megafauna reproduces slowly.

However, there are related factors that can affect the loss of species, and as an aid to analysis, we group them into two additional categories (Hurles et al. 2003). These are, first, habitat modification (including the use of fire, especially by early humans) and, second, the introduction (deliberate or accidental) of new plants, animals, and diseases—these can be considered, like humans, to be invasive species. Thus at least three categories of mechanisms are important. These three (hunting of megafauna, habitat modification, and the introduction of other plants, animals, and diseases) are each biological. All three can contribute to the loss of the megafauna, particularly if it is already under stress from major climate changes. However (unfortunately), we also expect other factors, including human-induced (anthropogenic) effects on the increase in CO_2 (and other greenhouse gases) leading to physical effects through climate change. Ongoing factors will include the expansion of the human population, with a larger and larger share of global productivity being devoted to human use. We consider these as 'biological' factors because they occur as a direct result of human activity; and they are, in principle, avoidable by modified human behavior, including the use of plastics. What is the role of new groups of taxa (including humans)?

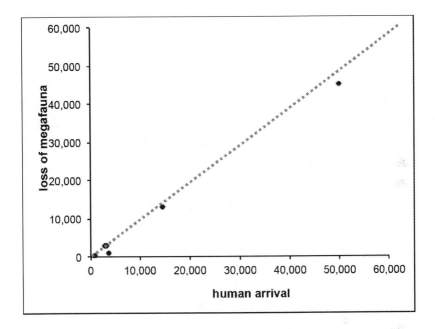

Fig. 6.2. Human arrivals versus times of extinction of megafauna. Under the human impact model, the changes will be after permanent settlement, although some invasive plants and animals may appear earlier. In practice, impacts on a naive fauna are simultaneous with or after human arrival. Effects might be slower on larger and/or more diverse landmasses; conversely, they should be faster in smaller more homogenous islands. Both axes are years before present (BP).

Many Earlier Ice Ages

There is a suggestion that there were some extinctions for plants at the onset of the first glaciation in North America (Figure 6.3), but that if the plants had effective dispersal mechanisms, then they would move north or south during the glacial cycles. This plant example is really as expected; most extinction would have occurred at the onset of the first glaciation and not very much later (as happened with the megafauna). Within the same time period (the last ~2.6 million years), there were also many additional smaller changes in average temperature.

So, and in a very fundamental way, there were dozens of major changes in climatic and temperature conditions, without any 'mass extinction'— until after modern humans arrived. The megafauna as a whole (and most other biota) survived all these major changes in temperature, possibly by moving north or south on more level regions (or in the marine realm) or to higher or lower elevations in steeper regions. As we have seen, the figure really eliminates climate change (by itself) as the main initiating factor for mass extinction of megafauna; the fauna had survived nearly eighty major changes in physical conditions. There will be other cases where there is normal turnover of species before humans arrived in (for example, Australia), but not mass extinctions. Thus major temperature/climate changes are not viable explanations for the ongoing sixth mass extinction. As referred to (particularly Figures 6.2 and 6.3), a major change in physical factors is *not* a guarantee (given species dispersability) that there will be major changes to the biota— above and beyond the usual turnover of taxa from normal evolutionary processes. So although we tend to eliminate climate changes (such as Ice Ages) as the initiating factors for the sixth mass extinction, they are clearly important.

Extinctions after Human Arrival

The second major point is that it appears that humans had always arrived before extinction of the megafauna. Thus the timing of human arrival and of extinctions (see earlier Figure 6.2) indicates that for several parts of the world, the local mass extinctions were subsequent to human arrival, and it follows that human settlement must be established before the full impacts become apparent. Australia is the first example, with human arrival maybe 50,000 years before present, and the extinctions continue until around 40,000 years before present, including as far south as Tasmania (Grun et al. 2010; Rule et al. 2012). Bird et al. (2008) also show habitat modification by the use of fire, and Lopes dos Santos et al. (2013) suggest that there are changes in the vegetation following the extinctions. Certainly, habitats will be affected differently, depending, for example, on other factors, including climate. Extinctions of megafauna in Australia are difficult to pin on climate changes 100–40 years before present, when the coincident cooling/drying trends expanded the more open habitats (Prideaux et al. 2009). However, the main point is that the megafauna were present at the time of human arrival. Again, the lack of 'mass extinctions' of the Australian megafauna during many earlier Ice Ages is an important control value.

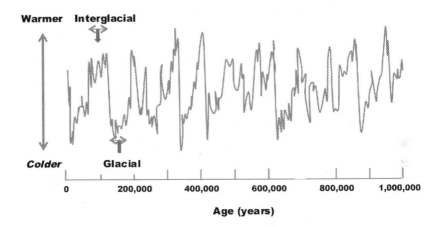

Fig. 6.3. The temperature changes during the last million years. The series starts with 41 thousand-year cycles (not shown) and later moved to the ~100 thousand-year cycles. There are c.38 Ice Ages, and so basically, the 'climate-change' hypothesis for megafaunal disappearance has failed over seventy times. It is essential to have these 'control' values when there is a major change in physical factors, but without a 'mass extinction'.

The next point on Figure 6.2 is the arrival of humans into North America and their eventual spread southwards to South America. The precise arrival time of humans into North America is still uncertain, but it is almost certainly before 14,000 years before present (see Gill et al. 2009; Johnson 2009). Gill et al. and Johnson also give extensive evidence to habitat modification as the result of human activity. So again, the effects of hunting, habitat modification, and the introduction of other plants and animals must all be considered.

The final examples come from the Austronesian (including Polynesian) expansion into regions such as the remote regions of New Caledonia, Fiji, Madagascar, and finally to New Zealand. For New Caledonia, Anderson et al. (2010) estimate human arrival by 3,000 years before present, and this is followed by the extinction of a giant megapode (a bird) and least nineteen other taxa, including a crocodile and turtles. White et al. (2010) report large horned turtles from cemeteries and middens in Vanuatu from around 3,000 years before present, again indicating that that component of the megafauna survived until after human arrival. There are extensive reports of now-extinct animals in Fiji from after human arrival (Worthy and Clark 2009). The Austronesians also reached Madagascar; and the megafauna includes giant lemurs, pygmy hippos, elephant birds, giant tortoises, etc. The precise time of settlement is uncertain, but still relatively recent compared with Australia or North America and is still part of the Austronesian expansion. Dewar et al. (2013) propose a settlement in northwest Madagascar ~4,000 years before present. Researchers assume a fairly small founding population, and consequently (given the relatively large land mass of Madagascar), it would have taken longer in some areas before the human population size increased to have significant populations over the whole country. However, the authors give evidence that the megafauna was still extant when modern humans arrived; some persisted (in some areas) until nearly 500 years before present. So again, an important point is that the larger vertebrates were still there at the time of the original human arrival.

The final (most recent) point on Figure 6.2 is New Zealand, where human settlement is estimated to be within the last 800 years (though discovery could have been somewhat earlier). The most notable extinctions were the giant ratite birds (moa), and here direct hunting (as well as the use of the eggs) appears significant. Large flightless birds (moa) are very likely to have used regular paths, thus making hunting them easier—it may have been only necessary to wait quietly, and dinner would walk by! It is important that there was no decline of the local megafauna before human arrival (Allentoft et al. 2014), and in this case, the effect of humans was relatively fast. However, there are expected to be both habitat modification and effects from the introduction of other plants and animals (Hurles et al. 2003). For the former effect, it has been known for over fifty years that extensive fires followed human arrival on the east coast of the South Island of New Zealand.

Separately from this, there have also been large numbers (probably over 1,000) of extinctions of birds on the Pacific Islands following human arrival (Duncan et al. 2013). The extinctions on Henderson Island are especially interesting in that it appears that birds were hunted and cooked more or less in an order of preference—middens have been dated and show a progression of the bird bones present (Wragg 1995). This island was later abandoned following the extinctions of many birds, and the island was not occupied by humans at the time of its European discovery. Other examples could be given—for example, the arrival of humans on the Caribbean Islands.

The results of Figures 6.2 and 6.3 are complementary—the second showing many major climate changes from the Ice Ages without extinctions of megafauna, and Figure 6.2 shows an association of the timing of extinctions following the arrival of humans. In all these examples, there is good evidence available that the various components of the megafauna were present at the time of human arrival. However, it is important to remember that the effects of humans include at least three classes of effects, direct hunting, habitat modification, and the arrival of other invasive species, including commensals. Geographically, from Madagascar to Australia to the Pacific to the Americas, the effects are dramatic. The results need to be considered as a whole, and no one region (for example, North America) should be considered separately.

Other Models

Other potential alternative hypotheses have been proposed; perhaps the most 'interesting' has a local asteroid impact, apparently sometimes called the 'fire and brimstone model'. It is interesting that even at the time it was proposed (Firestone et al. 2007), it had already been disproved as a general explanation because the timing of the extinctions at different places around the world varied from around 45,000 years before present for Australia (Grun et al. 2010) to only around 700 years before present for New Zealand (see Figure 6.2).

In other words, the local asteroid impact fails badly in its generality (we refer to it as the 'intelligent meso-asteroid destruction' (iMAD) hypothesis), and the relevant timings were known long before the original impact hypothesis was ever proposed for North America. It apparently envisages several asteroids circling the earth and with notes to the effect that the asteroids should (a) have a note as to which region they will strike, (b) wait for humans to arrive before striking the earth in that region, (c) kill the megafauna, and (d) under no circumstances are they to kill off the humans. This point has been emphasized because it illustrates the importance of nonscientific 'beliefs'; scientists are human—dammit! Fortunately, the asteroid impact during the Younger Dryas of North America has been refuted by many authors (see Sanchez et al. 2014), though a few authors still try to assemble some indirect evidence for the impact model.

From a scientific viewpoint, we prefer one hypothesis that can explain a wide range of data over a range of ad hoc hypotheses where each relates only to a single event—extinctions following modern human arrival in North America in this case. Anyway, Figures 6.2 and 6.3 illustrate the failure of the 'it must be a physical change' model—these figures are important control values. So for the last (sixth) mass extinction, we are left with the three classes of biological causes (direct predation, habitat change, and the arrival of invasive species). However, we stress that the human effect on extinctions is ongoing and depends, for example, on (preventable) climate change. However, the main point from this section is that the arrival of new species can explain the sixth mass extinction. But is it an asteroid impact for the fifth?

The Fifth Mass Extinction

We have now (finally) get to the real point of this chapter. The fifth mass extinction is around the Cretaceous–Paleogene (K–Pg)) boundary, now dated at about 66 million years ago. We have long suggested that there were more interesting biological complexities than are usually not considered. However, that is immaterial; we need the data. There are probably at least five important issues.

1. The impact was real (though so was the Deccan volcanism), but given the number of times that major physical changes have not resulted in worldwide mass extinctions, we do need good data on the worldwide biological consequences of the impact. These negative results are important control values.

2. The smaller dinosaurs were apparently decreasing in frequency before the impact (see Figures 6.4 and 6.5 and Box 6.1). Smaller terrestrial carnivores and omnivores were apparently absent by the Maastrichtian period (the surviving alvarezsaurids were apparently semifossorial).

3. We need to study much more the effect of population size of modern populations; the general understanding is that this is an important factor. Smaller dinosaurs (and smaller organisms with their larger population size) are the source of the future larger dinosaurs. Therefore, in the longer term, the disappearance of smaller dinosaurs is very interesting from a potential evolutionary point of view.

4. Studies from modern species imply there would be major competition between the early placental mammals and the smaller dinosaurs (and which is apparently what we see—the smaller dinosaurs are disappearing first). We need to study the interactions of placental mammals and reptiles.

5. Since writing this, it has been reported by Sakamoto et al. (2016) that dinosaurs were declining many millions of years before their final demise. This paper does not distinguish between larger and smaller dinosaurs but supports the conclusion here that dinosaurs were facing ongoing competition from placental mammals. Did placental mammals replace dinosaurs?

An underlying question then is whether the replacement by mammals of dinosaurs was a normal evolutionary replacement. Another way of making this statement is to ask whether the replacement of dinosaurs by placental mammals have occurred anyway regardless of the impact. This is an important question. There are a range of several fundamental questions here, and perhaps it helps to consider them as ranging along three dimensions of responses:

> Would dinosaurs still be present without the impact?
> What were the regional effects of the K–Pg impact?
> What were the effects on each of the dinosaurs, pterosaurs, and marine reptiles?

We do need to have good evidence and consider all the factors that might be important.

1. The realness of the impact.

It should be clear that the K–Pg impact is both real (and devastating, especially in the Caribbean (Schulte et al. 2010)). These authors report that deposits from the impact might be from 80–100 meters thick close to the impact site, decreasing to 2–10 centimeters thick in western interior North America and to about 2–5 millimeters in the rest of the world (>5,000 kilometers from the impact site). Was that 2–5 millimeters of deposition sufficient to eliminate the full range of many 'dinosaur' groups that had been abundant for millions of years? We certainly expect significant regional extinctions in the Caribbean and probably in much of North America. Claramunt and Cracraft (2015) suggest that many modern bird species survived that impact in South America. The original Alvarez et al. (1980) publication reported the iridium layer (that marks the impact zone) from two widely separated sites in Europe (Denmark and central Italy), together with the South Island of New Zealand. We can scarcely get more widely dispersed sites on the globe than these three, so there has never really been any question that the impact was real and had global effects. There is now good evidence for a major tsunami in the Caribbean caused by the impact, an impact site near Chicxulub in the Yucatan Peninsula has been identified, and there are over 350 sites around the world where the iridium anomaly has been identified.

The extraterrestrial impact was not the only factor towards the end of the Cretaceous period. The Deccan Traps volcanism in India has long been advocated as having an effect, and recent work still reports effects of the massive volcanism (Schoene et al. 2015). Indeed, it has been suggested that there may be no mass extinctions from any large single factor, but that maybe a combination is required to lead to a mass extinction. So the combination of events is certainly interesting, but it is not necessarily 'sudden and unexpected' extinction. Although we know that there will be many effects of local climate change (and other physical effects) over the many millions of years that multicellular animals have existed, this does not make any physical factor the automatic default explanation for any mass extinction; we must also consider new groups of taxa.

However, at the same time, as there is good data for both an extraterrestrial impact and for Deccan volcanism, there was no evidence presented in the original Alvarez publication about the biological effects of the impact. It was just assumed (without reference to any biological/ paleozoological literature) that there was an abrupt ('sudden and unexpected') mass extinction of pterosaurs, dinosaurs, and marine reptiles (ichthyosaurs, mosasaurs, and plesiosaurs). However, it appears that it is now accepted by many authors (though not by Schulte et al. 2010) that pterosaurs and some marine reptiles (such as plesiosaurs) went extinct for other reasons (not the effects of the impact nor of Deccan volcanism), leaving just the dinosaurs to (conveniently?) become extinct at the K–Pg boundary. This appears to be a significant difference from the original hypothesis that all five groups became extinct as a result of the impact, so the whole issue does need a full and detailed analysis, and we come back to this in the discussion of this chapter.

There is an interesting position that a number of authors appear to have taken recently with regard to the pterosaurs—that is, they (the pterosaurs) may have gone extinct naturally and, as judged by fossil footprints, through competition with early birds (Lockley and Rainforth 2002). But the extraterrestrial impact led to the extinction of the older lineage of enantiornithine birds (Longrich et al. 2011). So we do need some worldwide data comparing enantiornithine and ordinary birds (euornithines). We would need some reason why enantiornithine and not ordinary birds went extinct. Possibly if enantiornithines were more arboreal and ordinary birds more terrestrial (and given that trees may have been affected, as judged by a 'fern spike'), then this might be a reason. So we do require a careful worldwide study (of both avian fossils and the 'fern spike') close to the K–Pg boundary and especially outside of North America. This certainly merits more attention.

2. The decline of smaller dinosaurs (and pterosaurs)

Not only is there a major decline of dinosaurs generally (Sakamoto et al. 2016), but there is also a decline in smaller dinosaurs during the late Cretaceous period (Figures 6.4 and 6.5 and Box 6.1). It is particularly the decline in smaller dinosaurs that I concentrate on here because (as I explain later) they are expected to the source of later dinosaurs. And for pterosaurs, we have already shown (Slack et al. 2006) that there was an earlier and similar decline (and disappearance) of smaller pterosaurs. These were increasing in size quite dramatically (as measured by wingspan) during the Late Cretaceous period, but simultaneously the smaller pterosaurs disappeared—being apparently replaced in their niche (as judged by fossil footprints (Lockley and Rainforth 2002)) by sea- and/or shorebirds.

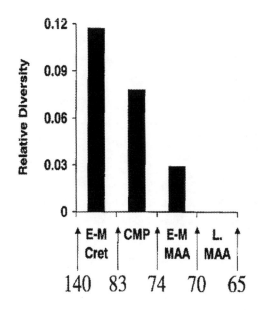

Fig. 6.4. Relative diversity of small dinosaur genera during the Cretaceous period, as a proportion of overall dinosaur generic diversity during the time periods: pre-Campanian Cretaceous (E–M Cret), Campanian (Cmp), and Maastrichtian (Maa). Small dinosaur genera were defined as those including species of estimated adult length less than 2 meters. The Maastrichtian data is very much influenced by North American information.

There is no question that the consequences of the asteroid impact was devastating (in the Caribbean at least), but the question is whether it unexpectedly/abruptly eliminated dinosaurs and pterosaurs worldwide. Under that hypothesis, there is expected be no real decline of pterosaurs and dinosaurs and no expansion of mammals and birds before the impact. As stated previously, in this sense, the changeover would be 'sudden and unexpected'. The Alvarez paper provided no evidence about when modern birds and mammals started diversifying nor when smaller dinosaurs and pterosaurs started declining.

Initially it appeared that there was a 3-meter gap between the last recorded dinosaur in North America and the iridium layer that marked the K–Pg boundary. More recently there has report of a single large dinosaur about 13 centimeters below the impact, though this does not affect the ratio of large to small dinosaurs (Figure 6.4). We will see in a later section that it is precisely the smaller dinosaurs that we would expect to lose first if there is competition between the early placentals with the dinosaurs. Perhaps the main point here is that additional work and/or calculations are definitely required.

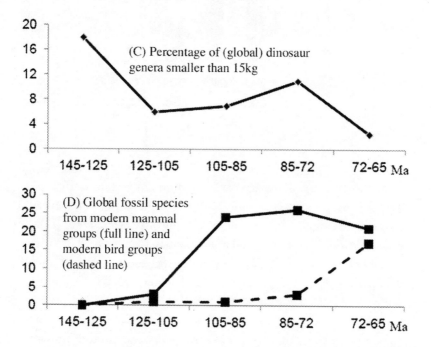

Fig. 6.5. (A) Percentage of smaller dinosaurs through time. **(B)** Increasing diversity of mammals and neornithes through time (based on Phillips et al. 2014).

3. Effect of population size

So this brings us to the effect of size (particularly population size), and again there are two aspects. The first is more 'descriptive' of what happened (many organisms get larger through geological time), and the second is more explanatory in terms of population sizes. As we have seen, it is much easier for larger populations of smaller animals to acquire complex adaptations (that might require at least two mutations, both neutral in themselves, but positive when combined (see Lynch and Abegg 2010); this is a standard picture from population genetics). This is because larger populations show more diversity, and it appears to be what is happening in practice. It is the smaller dinosaurs that are being affected first, and this is what we observe; the smaller dinosaurs (Figures 6.4 and 6.5) and also the smaller pterosaurs (Slack et al. 2006) are disappearing first. So this is what we would predict from a mechanistic point of view if the effects are basically responses to the diversifying placentals and birds. It is certainly a hypothesis that merits serious investigation.

	E Cret	A-A	C-S	CAMP	MAAS
	Omnivores				
Small (<20kg)	1	0	0	1	0
Large (>20kg)	5	3	4	6	8
	Carnivores				
Small (<20kg)	10	2	4	9	0
Large (>20kg)	18	22	19	24	17
	Herbivores				
Small (<20kg)	3	3	2	4	2
Large (>20kg)	40	53	54	69	51
	total				
Small (<20kg)	14	5	6	14	2
Large (>20kg)	63	78	77	99	76

Box 6.1. The apparent loss of small carnivorous and small omnivorous dinosaurs in the Maastrichtian period: Early Cretaceous (*c.*145–125 million years ago), A–A (Aptian–Albian, *c.*125–91 million years ago), C–S (Cenomanian–Santonian, *c.*91–83 million years ago), Campanian (*c.*83–71 million years ago), Maastrichtian (*c.*71–66 million years ago), (from Phillips et al. 2014 and the *Paleobiology* database).

To return to the first aspect (increase in size) is an example of Cope's rule (Hone and Benton 2005), which is the tendency for many organisms (animal and plant) to become larger through geological time. Cope's rule is usually treated as a descriptive/observational rule of what is observed, but the process only works (organisms getting larger) if there is continued loss/turnover of larger organisms. It is likely that both population size (Lynch and Abegg 2010) and long generation times (Zhong et al. 2014) are both important contributors to the mechanisms behind the processes. The first recognizably tyrannosaurid dinosaurs were about ninety times smaller (by weight) than the later giant tyrannosaurids, so this is consistent with a tendency to increase in size—but it has the inevitable tendency to smaller population sizes, with the inevitable(!) influence on long-term diversification.

Thus it appears to be a general phenomenon; larger dinosaurs begin as smaller dinosaurs and increase in size over time (Sereno et al. 2009). Population genetic theory predicts consequences from this effect; population sizes are expected to be smaller for very large animals, and small populations are much slower in developing new complex adaptations. Lynch and Abegg calculate, for example, the time for a more complex adaptation (requiring at least two mutations) to occur and to be fixed in the population. They report a major increase in times for such innovations for smaller population sizes. The effect is a major one, and so there are important limitations for complex adaptations requiring a number of mutations for smaller populations (and we expect that larger organisms will have smaller populations). So from a biological viewpoint, there do appear to be

> some important shorter-term advantages in being larger (it has happened many times in both plants (trees) and animals), but
>
> this increased size has a cost in having smaller population sizes that will make it more difficult for continued longer-term adaptation, and
>
> there is an effect of reduced mutation rate for long generation times (Zhong et al. 2014), and this will also markedly increase the time for complex new adaptations to arise.

However, more work is certainly required here. There appears to be both an advantage in having larger body sizes, but simultaneously a cost in having smaller population sizes and a lower mutation rate. This appears to be a generalization that works for both animals and plants (trees)! And it appears also that there is a faster turnover (shorter duration) of large mammalian species.

4. Interactions between reptiles and mammals

What models of competition between reptiles and mammals do we have available today? Do we find any mechanisms that we can study in the present that will help explain earlier effects in the past? An important example is the competition between populations of *Sphenodon* (tuatara) and small rodents such as the Polynesian food rat, kiore (*Rattus exulans*), that were introduced into New Zealand by early Polynesian explorers. These small rodents, averaging 60–80 grams, displace tuatara, which weigh up to twenty times their adult weight (up to 1.5 kilograms in some populations). Although recruitment of juveniles to adults ceases, adult tuatara survive in the presence of kiore (though not in the presence of the larger *R. norvegicus* or *R. rattus*). It is not certain whether the effect on juveniles is indirect competition for similar resources and/or direct predation on small tuatara. Whatever the mechanism, removing the smaller rodents from islands allows juveniles to survive and become adults (Towns et al. 2007). There are also separate examples in New Zealand of the effect of mice on survival of other small lizards (but the lizards are still larger than the mice).

There is also a report of the effect of mice from Gough Island in the South Atlantic. Wanless et al. (2007) report that mice systematically attack and kill albatross chicks that are up to 300 times their own body weight. The mice can be only 35 grams, and the albatross chicks up to 8 kilograms. Thus size in itself is not the issue; it used to be thought that 'Late Cretaceous mammals were small, dinosaurs were large, so therefore any effects of mammals on dinosaurs was not an issue'. This position can certainly no longer be maintained without some good evidence; even smaller mammals can have a devastating effect on other (and larger) animals, especially on their juvenile stages. Several authors report predation by mammals on the young of dinosaurs. O'Gorman and Hone (2012) give a size distribution of many classes of vertebrates and show the skew on the large size of dinosaurs, which emphasizes that smaller mammals might be having an effect, especially on juveniles.

As mentioned earlier, a possible reason for the potential effect on larger lizards by smaller mammals is given by Pond (1977). She points out that today mammals and birds are the main land vertebrates that feed their young until the young are able to function in the same niche as their parents. As such, mammals and birds have only to optimize for a single niche, and other authors reiterate a similar point—that it is important for understanding basic physiological factors that are an aid in the survival of the offspring. So there are at least two major effects of feeding the young; it allows more specialization for a single niche, and it gives even more protection of the young. It is predicted from our current knowledge that the smaller mammals that existed before the end of the Cretaceous period (often up to 1–5 kilogram) would affect the smaller dinosaurs, which is what we observe. It is the smaller dinosaurs (and pterosaurs) that disappear first. This is important because it is the smaller dinosaurs that give rise to later (and larger) dinosaurs. This generalization of larger plants and animals really applies to a wide range, and we return to this aspect again in the discussion.

An obvious point here is that we cannot guarantee that early mammals (before the end of the Cretaceous period) had all the foraging abilities of the myomorph rodents (that includes rats and mice). Rodents only arose later during the Tertiary (Paleogene) period, but their place may have been taken earlier by different stem-group mammals—for example, the multituberculates, which began diversifying at least 20 million years before the end of the Cretaceous period (Wilson et al. 2012), about the same time as dinosaurs began to decline (Sakamoto et al. 2016). It is not relevant here that the multituberculates are now extinct—they certainly were living at the end of the Cretaceous period and probably until the Late Eocene period (about 35 million years ago). The change over time does not in any way invalidate conclusions of effects 66 million years ago. Archibald et al. (2011) report a *Protungulatum* fossil from the Late Cretaceous period of North America. Maybe *Protungulatum* is not within the current crown group of modern placentals, but it does show that ecological diversification of placentals predates the boundary. A recent reference suggests that *Purgatorius* was an early primate. A much older group of mammals (that includes *Necrolestes*) has recently been identified from South America, where it survived until less than 20 million years ago. Nevertheless, there apparently was predation by mammals on the young of dinosaurs.

Characteristics of placentals generally appear in the interaction of rodents with 'lizards'. We find similar effects of feral cats and/or ferrets on biota, indeed, it was only after the removal of cats on Gough Island that the mice numbers increased and the mice started attacking seabird chicks (Wanless et al. 2007). The main point is that this invalidates any claim that, because placental mammals were relatively small during that time, they would not have affected the dinosaurs during the Late Cretaceous period because the dinosaurs were larger. This suggestion is clearly not correct—and if early mammals, for example, did outcompete the smaller-sized dinosaurs (possibly by direct or indirect competition with the young of the smaller dinosaurs), then it would be expected that this would eventually have major long-term effects on the larger dinosaurs. Perhaps all that this suggests is the mammals would have eventually replaced dinosaurs, irrespective of the impact. It is possible to maintain that intermediate position even though the disappearance of small dinosaurs had long-term effects (see the final section of this chapter). We must always use all known mechanisms and test our hypotheses.

5. Recent work, the decline of dinosaurs generally

As I have already pointed out, since this was written, it has been apparent that dinosaurs were in decline for at least 25 million years before their final disappearance (Sakamoto et al. 2016). These authors report that dinosaurs were below the replacement level for a very long time before their final extinction about 66 million years ago. Their speciation rate fell during all that time. Okay, we do not know if they persisted somewhat longer in parts of the southern hemisphere. But the main point is that the rate of new species formation declined below the extinction rate, and there must be some reason for that.

This does not affect in any way the loss of the smaller dinosaurs first, but it reinforces the conclusion that dinosaurs died out naturally. Anyway, this is excellent support that the placental mammals replaced the dinosaurs.

Conclusions

So researchers currently favor very different mechanisms for the fifth and sixth mass extinctions—and we need to test our favorite models. In one sense, we expect that the conclusions about the arrival of humans before the 'mass extinctions' started to be fairly firm, so we should make faster progress by testing different aspects of the fifth mass extinction from 66 million years ago. Nevertheless, we need to be able to test all our models/ideas. And as we point out earlier, authors do not necessarily agree on the best explanations, though they should welcome additional testing.

First, though, it is probably important to clarify again what is *not* being said. The K–Pg impact was damaging, especially in the Caribbean and North America. It really was devastating, and deposits 80–100 meters thick close to the impact should remove any doubt. What we are implying is relatively simple: the development of new groups must always be considered when seeking to understand past processes; they are part of microevolutionary processes. We certainly expect that many factors will have important effects even if the effects are not mass extinctions and even if the effects are delayed by many centuries.

The problem is apparently not a new one. Ospovat (1981) describes some of the reasoning that Charles Darwin went through between 1844 and 1859 before he had decided from his extensive readings of the biological and paleontological literatures that all factors (including biological) were important during evolution. Attributing major changes in the fossil record exclusively to factors other than the development of new groups does not meet the goal of using *all* factors that can be studied or inferred. Certainly, Darwin adopted the Hutton/Lyell approach of explaining the past by processes that could be studied in the present (and Lyell at least considered extraterrestrial impacts). Nevertheless, it was an important development in evolutionary theory that Darwin did extend Lyell's approach to biology, though some authors suggest that Darwin had little major effect on some aspects of paleontological thought. However, our point is more about formally considering all mechanisms available for evolutionary change.

Perhaps the problem is 'gradualism/continuity' itself. Some paleontologists appear (to some at least) to be rejecting gradualism/actualism as a mechanism in favor of some system of more rapid and sporadic evolution by an unknown mechanism. But animals improve 'gradually' to an improved system (and it is affected by population size). Thus it appears that some people are rejecting a key part of Darwinian evolution.

It is not helpful to appeal to such factors as the 'incompleteness of the fossil record'. The real problem is a failure of science itself, failing to test all possible models critically. We all have a tendency to 'believe' our models and do not want to test them scientifically. We tell our undergraduates that 'belief is the curse of the thinking class'. In the last forty years, we have gone from models where there were major biological effects to those apparently only considering physical changes (but see also Lloyd et al. 2008). It is clear that the impact is real, and the fossil data is brilliant; it is just the interpretation that lacks testing.

Although we have concentrated on the most recent two mass extinctions, the principles may be the same for the earlier ones. Stigall (2012) has already suggested the role of new invasive groups in the extinctions towards the end of the Devonian period, both biotic and abiotic factors being important. There is report of an important stage during the Late Triassic period with an extended period (of up to 20 million years) with an intermediate assemblage of earlier and later dinosaurs coexisting. During this time, there appears to be a turnover of taxa (but still called 'dinosaurs'). This may have been an equivalent period to the Late Cretaceous times, during which early mammals and late dinosaurs competed. Fundamentally, we see the opportunity to study all factors that might help lead to mass extinctions as an opportunity for the future even though we know that climate changes will always be important—they are occurring nearly all the time, at least on a geological timescale. We see mass extinctions as an opportunity to test alternative hypotheses.

It is very likely that there were quite separate (nonscientific) reasons affecting the early reception of the extraterrestrial impact and its suggestion that the impact 'caused' mass extinctions. The time of the original Alvarez et al. (1980) publication was the height of the Cold War (with its apparent reliance on retaliatory attacks at any sign of nuclear weapons being used—mutually assured destruction (MAD)), and the effects of such an all-out escalation was a concern of many scientists. There was widespread concern at that time that an all-out nuclear war would lead to disasters such as a 'nuclear winter', with many clouds of dust getting into the high atmosphere and blocking out sunlight—leading to unpredictable, but very negative, effects. So in this sense, there was a positive effect of the Alvarez et al. publication, but it was not a scientific response, and that situation is (hopefully) now much improved. However, as scientists, we have to put all that aside and have to have to face the evidence available to us now about the consequences of the asteroid impact at the end of the Cretaceous period. Any nonscientific response, no matter how idealistic, is not relevant.

The future—we need to integrate more studies from many different disciplines. We cannot do without the brilliant fossils studied by many researchers and worldwide. But we also need a very mechanistic genetic approach to the methods—are the mechanisms that we can study in the present *sufficient* to explain past effects? The future is very positive; there are real advantages of a combination of interdisciplinary researchers working together on a problem. In principle, we see the necessity of integrating several lines of evidence to make real progress on testable predictions. But what would happen if we could bring back a group of dinosaurs to the modern world? Would they be able to compete with mammals at all stages of their life cycle? Why were the smaller dinosaurs (and dinosaurs generally) apparently disappearing in the very late Cretaceous period? Were the oviparous dinosaurs really able to hold their own against the viviparous mammals? Were the more meso-thermophile dinosaurs (Grady et al. 2014) able to hold out against the more endothermic mammals? However, nothing we have observed yet in the evolution of genomes contradicts that macroevolution is the same basic processes as microevolution but continued for longer times. But we must continue to test our models!

Regardless of the implications of the results, it is important to study the potential effects of the diversifying placental groups and their effect on earlier top carnivores and herbivores. Perhaps a real question to focus on is whether the non-avian dinosaurs would have gone extinct anyway (without the impact in the Caribbean), given the diversification of the birds and mammals. There are certainly mechanisms that would make this appear almost inevitable. What were the advantages of placentals and birds that they survived the K–Pg extinction event?

Whatever our viewpoint, it is absolutely essential to test our hypotheses. Although the original publication (Alvarez et al. 1980) also suggested the 'sudden and unexpected' model applied for marine reptiles and pterosaurs as well, these appear to have dropped out of the 'sudden' extinction model. We need to test our models—our 'beliefs' are not evidence. We need to understand the mechanisms of evolution and determine whether the past can be explained by current causes, including biological factors. Perhaps the simplest way of putting the basic question is to ask whether placentals are now more effective than dinosaurs in most land environments. Is there any 'progress' during long-term evolution?

CHAPTER 7

Which Came First, Eukaryotes or Akaryotes?

This second Popperian question certainly requires much more discussion than it usually gets, and it should be accepted that we just don't know the correct answer. At a fundamental level, there are two types of cells: eukaryotes (with a true nucleus) and akaryotes (or prokaryotes, without a nucleus, and I will use this more neutral terminology of akaryotes, as advocated by Patrick Forterre). It is usually assumed (without formal analysis) that prokaryotes arose first and eventually gave rise to eukaryotes, and that is implied in the name 'pro'karyotes. 'We' (that is, humans) were eukaryotes and therefore must surely be the most advanced organisms—appears to have been a reasoning (or lack of it), but that is not evidence. The bacteria are brilliant at what they do! We have already seen the results of Lynch and Abegg to the effect that smaller organisms (with larger population sizes) are much more likely to get combinations of mutations that may be innovative, such as in the biochemically outstanding bacteria.

There are two groups of akaryotes: bacteria and archaea. The 'root of life' (as we know it) is between these two akaryote groups. So this is one point to the standard idea that akaryotes come first: eukaryotes have to be formed only once (by some still unknown selection processes). However, for example, it seems that ancestrally all eukaryotes have their genes broken up into an exon/intron structure—the exons code for the final protein, and the introns are excised out from the mRNA. The primary catalysts are five RNA molecules (assisted by a large number of proteins). The usual explanation is that the five RNAs are (in some mysterious way) derived from RNA group II introns that exist in bacteria, but there is no evidence for this. But the standard explanation for the origin of proteins is that they are derived from peptides (coded for by exons). So then it is even—one point to akaryotes being the first and one point for eukaryotes being first! But which way did life go? Okay, we assume that (because all life has essentially virtually the same triplet code) all life is related—this is an important point.

That akaryotes come first is the usual answer, but maybe it is just a 'belief'. This question requires much more discussion—do the bacteria and archaea really precede eukaryote cells? Or are all those features found in Box 7.1 (see later) really ancestral? Thus from our Popperian perspective, we should see if there is any real evidence. We 'believe' that akaryotes are the earliest forms of cells and that eukaryotes (and all their properties) are in some unknown way derived from them. So it is really an important scientific question: do the earliest cells have both a genetic compartment (nucleus) and a metabolic compartment (cytoplasm), or do they already combine them into a single compartment, as akaryotes do? This chapter is about these problems.

We will first consider the nature of both the last universal common ancestor (LUCA) and the last eukaryotic common ancestor (LECA, sometimes called FRED, or 'fairly remote eukaryotic daddy'). They are very different organisms, but we can get them mixed up! LUCA is the last of the organism that is ancestral of all modern life forms: archaea, bacteria, and eukaryotes (or caryotes on the A, B, and C models, which we should probably use for this question). It must have had RNA in many forms and DNA and proteins and the standard code for relating mRNA to amino acids. It was just one stage in the origin of life. It is assumed that it has the standard code for amino acids.

LECA (FRED), on the other hand, refers to just the last eukaryotic common ancestor (of extant (living) eukaryotes only—a much more restricted group. We should recall the standard picture of 'stem groups' and 'crown groups' that is outlined earlier in Chapter 6 (Figure 6.1, and it was illustrated by the five or six main groups of eukaryotes). The 'crown' group of eukaryotes are all related in that they are descended from the last common eukaryote ancestor (LECA/FRED). But there would have been earlier groups of eukaryotes that are now classified as 'stem' group eukaryotes, but which have no living descendants. The combined two groups are the 'total' group. Of course, we do not have direct knowledge of the earlier forms of eukaryotes that form the 'stem group', though we can infer a lot about their properties, and we address this question next. Based on the properties that appear to be common to most of the main groups of eukaryotes, LECA appears to have the genetics required (Box 7.1) for meiosis, mitosis, nucleus, nuclear substructures, an exon/intron gene structure, spliceosomes (with their many RNA and protein molecules), many centers of DNA replication, etc. (and including mitochondria). It is agreed that mitochondria are bacterial endosymbionts, having a bacterial ancestor but currently living entirely within the eukaryotic cell, and to have lost many of their genes to the nucleus. It is often assumed that early cells were akaryotes, and this is discussed later, but it gives us no idea about the nature of the LUCA. Anyway, even if the ancestral life form were more eukaryote-like, LUCA and LECA (FRED) are very different organisms.

Box 7.1. Some properties inferred to be in LECA
(the Last Eukaryotic Common Ancestor)

Nucleus/cytoplasm subdivision   ✓
Nuclear substructures (nucleolus, etc.)   ✓
Larger genome size   ✓
Exon/Intron structure of genes   ✓
Spliceosomal proteins with RNA catalysis   ✓
Capping of mRNA   ✓
Mitosis and meiosis   ✓
Centrioles and centrosomes   ✓
Linear chromosomes and histones   ✓
Many centers of DNA replication   ✓
Apoptosis/argonautes/dicer   ✓
Endoplasmic reticulum (and substructures)   ✓
Membrane lipids   ✓
Epigenetics   ✓
Flagella/Cilia   ✓
Mitochondria (endosymbionts)   ✓

We certainly do not expect all features to be in all eukaryote cells. For example, it appears that bdelloid rotifers have lost meiosis but perhaps compensate for this loss by the increased ability to take up genes from other organisms by lateral gene transfer. However, features such as the loss of meiosis in bdelloid rotifers is very unlikely to be the ancestral state for eukaryotes, so it does not affect the generalization that it is most unlikely that any of the main features will have been gained independently on all the five or six main lineages of eukaryotes, leading to Box 7.1. To allow for this loss of some features in specific groups, we say that we are seeking general (widespread, but not 'universal') features of eukaryotes. So if a feature is in each of the main groups of eukaryotes, then by far the simplest explanation is that it was also in LECA.

The usual hypothesis is that eukaryotes arose from some sort of fusion between an archaeal host cell and a bacterium, which became the bacterial endosymbiont (the mitochondrion). However, this hypothesis (by itself) does not explain the many features of eukaryotes, for example, the multiple centers of replication, the RNA-based elimination of exons in the mRNA, the nucleus/cytoplasm distinction, etc. In general, these features of eukaryotes are not at all well explained by the fusion of an archaeon and a bacterium (see also Forterre 2013). We use the term 'protoeukaryote' for a hypothetical stem group ancestral eukaryote that took up a bacterium as an endosymbiont that formed mitochondrion. Originally we used to use the term 'archezoan' instead of 'protoeukaryote', but this term was widely misunderstood. We had split the archezoan concept into two: the complex nature of the original ancestor of eukaryotes (which we agreed with) and that some current eukaryotes had always lacked mitochondria (which we disagreed with). But most others preferred the single concept of some existing eukaryotes never having had mitochondria, so we desisted and now use the protoeukaryote concept instead. There are some eukaryotes that live in an oxygen-free environment that have 'mitosomes' or 'hydrogenosomes' that are apparently ultimately derived from mitochondria. At least the term 'protoeukaryote' is understood to imply an already complex ancestor that took up the mitochondrion.

A related question is the timing of the most recent common ancestor of the eukaryote crown group (LECA), and it is estimated to be around 1.75 (1.87 to 1.68) billion years ago. We do not know the age of the 'stem' group eukaryotes, but our later discussion implies that they are very old. Some very large cells (>300 µm in diameter compared with 3–5µm of most bacterial cells) occurred around 3.2 billion years ago (Javaux et al. 2010), and they are possibly 'stem group' eukaryotes. However, they do not appear to have a nuclear region, so the possible age of the stem group eukaryotes is an important issue for the future. However, it is a real problem that we do not know the age of LUCA; we have little idea of its age, but it must have been many billions of years ago. We see possible organisms about 3.5 billion years ago, but we have no real idea whether they have the full triplet code like modern organisms. There are also several models that suggest that DNA evolves separately in different lineages after proteins arose—again there is no evidence for this (but that does not stop ideas).

As an aside, it is possible that the archaea were initially an adaptation to life at higher temperatures? Their lipids are very different from those in bacteria and eukaryotes and appear to be able to form membranes at higher temperatures. Their lipids are primarily isoprenoid compounds (based on a five-carbon branched side chain) in contrast to the more standard phospholipids of bacteria and eukaryotes, which are therefore more common. Similarly, the phospholipid has the three-carbon compound glycerol, and they are ester linked. In contrast, the archaea have ether-linked lipids. We do not know the temperature at which life arose, but a lower temperature origin would explain this anomaly by suggesting that archaea have adapted to higher temperatures. This would explain bacteria and eukaryotes having relatively similar membrane structures and that archaea have derived membranes—this possibility has to be evaluated. We also need to know much more about the adaptation of some archaea to include growth at higher temperatures (Boussau et al. 2008) and their use of some different cofactors (and their temperature stability). We assume that there were intermediate stages!

It appears easier to make detailed models of the origin of protein synthesis with a eukaryote to an akaryote transition rather than vice versa, but that is not real evidence. Thus we really are at a phylogenetic impasse in not being confident about the direction of change between eukaryotes and akaryotes. Let's look at Box 7.1 again; it shows some of the increasing number of properties that appear to be in the LECA (the current crown group). Although the protein complement of LECA is very important, we find it more useful to focus on the main cytological features rather than on the proteins as such (but both the proteins and the cytological features are important). However, at least since Reanney (1974), it has been questioned which direction the change occurred—akaryotes to eukaryotes, eukaryotes to akaryotes, or somehow are both forms equally old?

The question as to whether eukaryotes are oldest has been raised quite a few times, but the 'protoeukaryote' question has never been satisfactorily addressed by advocates of the standard 'akaryotes old' viewpoint. And they might be correct, but we need to evaluate all the ideas scientifically. Eukaryote phylogeny is still uncertain at the deepest (oldest) levels, and we have already discussed the limitation of obtaining a clear phylogeny of eukaryotes is that the mathematics of Markov models limits the distance back in time that we can reasonably and confidently infer phylogeny from sequences—but biologists love to ignore these problems. At the deepest times, the Markov models we currently use lose information exponentially even though there is also a linear increase in information as the number of sequences increases. The mathematical result of Mossel and Steel (2004) does *not* mean that there is no information about the deepest lineages—just that the Markov models we use for sequence data must lose information; there may be other information in the sequences. We try all sorts of weighting measures to try and recover deeper eukaryote relationships. One approach that we use is to infer ancestral sequences and then to use these ancestral sequences for further searches (Collins and Penny 2005). Another uses the inferred three-dimensional structure of proteins (e.g. Daly et al. 2013).

Perhaps the best that we can currently do is to identify five (or six) main subgroups of eukaryotes. An example of the problems that may occur is explained in Embley and Martin (2006) when they report that certain parasitic genera have accelerated rates of evolution, and some of these parasitic genera appear to have lost aerobic mitochondrial respiration (they live in anaerobic environments). Given these difficulties, our approach has been to find features that are in all of these main subgroups of the current (crown group) eukaryotes and so then to infer that they were also in the last common ancestor of eukaryotes, or LECA.

There are currently considered to be five or six main groups of eukaryotes; some people allow some more. These are at least the following:

> Fungamals (animals, fungi, and a few related organisms often called 'opisthokonts' because very few people know what that term means!)
> Amoebozoa

Excavata (including *Giardia*, *Trichomonas*, and *Euglena*)
Archaeplastida or plantae (red and green algae and land plants)
SAR supergroup (stramenopiles, alveoloates, radiolarians) and
  chromalveolates

There, you didn't really want to know that! Many people put the first
two together. But there are still other minor groups whose position is
uncertain; they appear in different phylogenies. Several authors have
their favorite view of eukaryote relationships, but the main point is
that we do not (yet) know the deep phylogeny (including the branching
order) of the current eukaryotes. Mathematically we expect that
additional genomes will help, but at present even the position of the
root of extant (living) eukaryotes is uncertain; some authors think
that it is within the Excavata, and others put it between the fungamals
(opisthokonts) and everything else. However, many biologists want to
go even deeper and determine archaeal or even bacterial phylogeny.
They could even be right; akaryote proteins might be easier to align
than eukaryote proteins. Eukaryote proteins are often longer and have
expansion segments within them (that perhaps should be deleted before
alignment). But we also need a thorough study of the enzymes that
considers the loss of information at deeper divergences.

We will continue with the background information for a little while.
Unfortunately, there is nothing in Mossel and Steel's results that limit their
conclusions to eukaryotes, and so we cannot be confident about the deeper
phylogenies of the archaea nor the deep aspects of bacterial phylogeny.
Some authors now suggest the eukaryotes arise *within* the archaea (making
the archaea polyphyletic). However, these trees are rather weak and are well
outside the time we expect phylogenies to be fully accurate (but that does
not stop some advocates). At the present (and until the issues are really
addressed and resolved), we should just consider the archaea, bacteria, and
eukaryotes as separate groups—and this leads to good hypotheses about
the three groups. Anyway, given the lack of confidence about the deeper
aspect of eukaryote phylogeny, our general approach is inferring properties
that are common to all main groups of eukaryotes.

However, the main issue is the common features of eukaryotes. Box 7.1 includes the main distinguishing cytological features of eukaryotes, each of which has many enzymes associated with them. For example, if each of the main groups has capping of mRNA, then the simplest hypothesis is that it was the common ancestor (LECA) that also had capping of mRNA. The structure of the eukaryote chromosomes are complex, and it is not a valid evolutionary argument to 'just assume' that these features arose somehow from akaryotes. Box 7.1 is critical to our understanding of eukaryotes; its importance cannot be overestimated in this regard.

Our discussion does mean that it is not clear yet whether the LUCA had more eukaryote or more akaryote properties or that it even matched one of these groups. It is usual to infer an akaryote LUCA, but as Figure 7.1 shows, it depends on the nature of the outgroup (and you can place crown groups to your heart's content on Figure 7.1C). Perhaps we also favor a lower temperature origin of life partly because the entropy (order) component of free energy ('-TΔS') is increasingly less optimal (or more difficult) as the temperature rises, and similarly RNA is increasingly less stable as the temperature increases; the backbone breaks, and cytidine (in particular) is unstable. Certainly, some recent work has favored a lower temperature origin of life (e.g. Attwater et al. 2013), and this does support models such as the akaryotes being derived from eukaryotes by increased resistance to temperature (Forterre 1995, see later).

There are many important differences in substructures of eukaryote and akaryote cells, and we will consider hypotheses for the origins of some of these differences. There are not yet any known evolutionary mechanisms that would favor the evolution of all the features of the eukaryotic cell from akaryotes. In contrast, there are several genuine (mechanistic) evolutionary mechanisms that would favor the derivation of akaryotes from eukaryotes—akaryotes are very sophisticated cells (and many have very large population sizes, which supports their biochemical innovativeness (Lynch and Abegg 2010)). Nevertheless, we must keep an open mind and still search for mechanisms that would favor an akaryote to eukaryote transition, and that would explain all the features in Table 7.1.

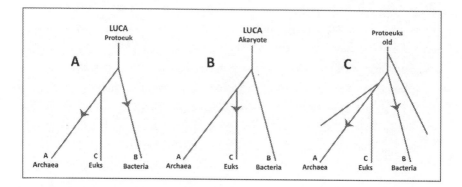

Fig. 7.1. The nature of LUCA depends on the nature of the outgroup. If the LUCA was proto-eukaryotic-like, (**A**) there are two changes to akaryotes required. If LUCA was akaryotic, (**B**) there is just a single change to eukaryote status (see Box 7.1). What was the nature of the LUCA? **C** shows that the concept of stem and crown groups does make a difference. This diagram assumes the standard arrangement of archaea, bacteria, and eukaryotes and should be read with the figure of 'crown' and 'stem' groups (see Figure 6.1).

Separation of Nuclear and Metabolic Compartments

An important question is the origin of the two main compartments of the eukaryotic cell. It is interesting that eukaryotes have membranes separating the genetic component (the nucleus) from general metabolism (the cytoplasm), even if the separation usually breaks down during cell division and reforms later. In akaryotes both the genetic compartment and metabolic compartments are very effectively combined with no membranes separating them, though there are some other subdivisions within some akaryotes and other membrane-bound compartments in eukaryotes (see Diekmann and Pereira-Leal 2013). The two cell substructures are shown in Figure 7.2 and help us on this important question.

Is there any reason why the single compartment (akaryotes), or the two compartments (eukaryotes), is more 'advanced' than the other? Maybe we should be more specific—is there any advantage of going from a single compartment to two or vice versa? Perhaps the size of the cell is part of an explanation. But we need a real answer, not a 'just so' story. One option for the reverse direction, deriving the single component from the two component system, is considered later in the section in 'Models for the Origin of Akaryotes from Eukaryotes'.

In addition, there are many substructures within the nucleus; for example, there is extensive localization/specialization within these nuclear regions, such as the nucleolus, the spliceosome, speckles, nuclear pores, capping of mRNA, etc. The nuclear pore is a very complex structure (that is made up of around thirty proteins (Von Appen et al. 2015)). What is the evolutionary advantage of each of these nuclear processes given that akaryotes manage very effectively without them? Are they early features dating back to the origin of life ('primitive features'), or are they somehow advanced features?

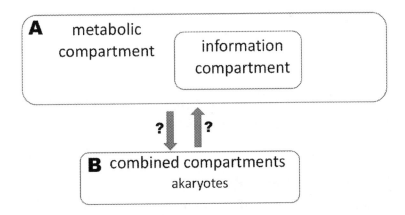

Fig. 7.2. The eukaryotic (A) and akaryotic (B) cellular structures. The two arrows indicate that we do not have evidence which was the original condition. Superficially we might expect the earliest cells to have the two differentiated compartments because this would fit in with the origin of proteins from short fragments (that is, having an exon/intron structure).

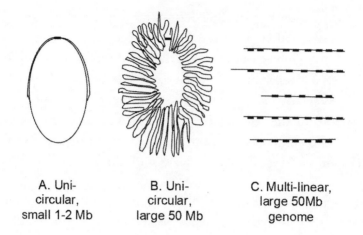

A. Uni-
circular,
small 1-2 Mb

B. Uni-
circular,
large 50 Mb

C. Multi-linear,
large 50Mb
genome

Fig. 7.3. How many centers of replication did the ancestral life forms have? Bacteria generally have a single center of DNA replication, though it is up to four in archaea. Eukaryotes have hundreds or thousands of centers of DNA replication (as in C). The time required to copy a very large genome (B) from a single center of replication (even copying both directions) would certainly be excessive (many hours). So we certainly expect that bacteria will be under strong selective pressure to minimize the size (length) of their genomes. But the original question remains, and single-stranded RNA-based viruses (because they have a high error rate) are expected to have had many centers of replication.

Effects of a Single Center of Replication

Another major difference is that bacteria usually have only a single center of DNA replication (though they may carry important genes on plasmids), whereas eukaryotes have hundreds to thousands of centers of replication over their many chromosomes (archaea may have one to four). It is a critical point in that the size of the genome is important for times of replication (see Figure 7.3); we certainly do not expect any very large genome to have just a single center of replication—it would take far too long to copy and correct the entire DNA from just a single center. So we certainly predict much smaller genome sizes within, say, bacteria; and we expect that there will be genome compaction (without large amounts of 'not really necessary' DNA). The time for replication is critical. *E. coli* may replicate every twenty minutes even though it takes longer to copy its DNA, so there must be mechanisms for keeping the different DNA copies separate (and attached to the outer membrane). However, this does not negate the principle that just a single center of replication will be limiting for the time of DNA replication. Whatever the mechanism is for DNA replication, we expect that organisms with a large DNA content and a single center of replication would take a very, very long time between cell divisions.

Indeed, the fastest cell (or at least nuclear) division that I know about is the nuclear division in *Drosophila* embryos—the embryo initially enlarges to form a very large cell, after which the nucleus (only) just repeatedly divides and divides about once in every eight to nine minutes. So this is faster than the generation time of a standard free-living *E. coli* cell. Thus the DNA (but not the whole cell) of eukaryotes can be copied quickly, given their many centers of DNA replication. This is not the usual way in eukaryotes, but it illustrates a point about the time to copy a genome.

It is not an evolutionary valid argument to say that eukaryotes 'needed' multiple centers of replication 'because' they were going to have larger genomes. There has to be an immediate short-term advantage for all the steps when increasing the genome size—and to go to multiple centers of DNA replication. It may be that the ancestral condition is multiple centers of replication based on the Eigen limit of how long pieces of RNA could still be copied without too much error (before going into error catastrophe). So once again, we are left with an impasse about the direction of change between akaryotes and eukaryotes in that there is not yet a good known mechanism for an akaryote-to-eukaryote transition for the number of DNA replication centers. Perhaps during the RNA and RNP worlds (before DNA), there was a stage with multiple centers of replication when each RNA molecule had to be copied more or less independently. So it is possible that the multiple centers of replication in eukaryotes could be more 'primitive' in that it might be derived from an early system of DNA replication, where the DNA was synthesized in small packages that stayed within the Eigen limit.

Splicing and the RNA Components of the Spliceosome

How did proteins arise? It was initially suggested (Gilbert 1978) that there were initially small 'peptides' coded for by introns and that evolution was possibly sped up by recombination to give additional early proteins (see Figure 7.4). Very early recombination allowing new combinations for proteins was important in this approach. In contrast, there are many theories that simply say something like 'and eukaryotes acquired introns—and that they arose from type-2 introns of bacteria'! But this does not give anything like an evolutionary mechanism of the selective pressures for how modern protein structure may have arisen. Currently the spliceosome has five RNA molecules and a large number of proteins that can now be found even in distantly related eukaryotes (Collins and Penny 2005), and the RNA components appear to catalyze the splicing reaction (Strobel 2013).

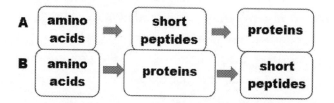

Fig. 7.4. Two models for the origin of proteins. Under A, there were earlier short peptides (coded for by introns), and then they were later joined by the spliceosome. Under model B (the standard model), proteins arose somehow and then much were later coded for by exons—the intron/exon structure is thus a later addition. There is usually no reasoning given for the origin of the catalytic RNA and the proteins that make up the spliceosome.

In general, proteins appear to have much faster catalytic reaction times than does RNA (Jeffares et al. 1998). So there is an interesting question about 'why' RNA has any catalytic functions at all—unless it was an ancestral condition. The faster reaction times of proteins are particularly important for reactions involving small molecules because the diffusion time for the two macromolecules to interact appears limiting. This might explain the observation that it is the processing of macromolecules where RNA tends to carry out the core reaction. Similarly, there does not appear to be any real immediate advantage in secondarily forming an exon/intron structure, together with the extremely complex spliceosome, once proteins have already arisen. Yes, there can be alternative splicing, but that still requires splicing to already exist—there is no evolutionary mechanism that splicing will evolve now 'because it will allow alternative splicing many millions of years into the future'. Any evolutionary mechanism must have an immediate short-term advantage, so some advantage of splicing must be in the short term. But once splicing exists, alternative splicing is a good possibility.

Introns can be added or subtracted to eukaryotic proteins, so it is a dynamic process. So proteins transferred to the nucleus from mitochondria or plastids, for example, can gain an intron/exon structure. Furthermore, some of the constraints on intron definitions appear to be highly conserved, even between brown algae and humans. Eukaryotes with the fastest life cycles tend to have fewer introns in proteins (Jeffares et al. 2006); it may take two to three minutes to excise an intron. This may be an explanation why intron loss seems more common than intron gain (Roy 2006). However, this does not answer the reason why eukaryotes have them in the first place and whether the spliceosome was present in the earliest living cells. It appears that some nucleomorphs (highly reduced secondary nuclei resulting from secondary endosymbiosis) have lost splicing altogether (Lane et al. 2007), so the spliceosome might have some other function/role in eukaryotes. So at the moment, the only real (mechanistic) evolutionary hypothesis we have is something like the original Gilbert-type (1978) one that some form of splicing occurred early and eventually provided a mechanism for the origin of proteins. One possible mechanism for the origin of proteins is given in Penny and Zhong (2015).

Meiosis and Mitosis (see Egel and Penny 2007; Wilkins and Holliday 2009)

Mitosis is an interesting aspect in that all cells must be able to divide, but we will not follow that up further here—both eukaryotes and akaryotes have cell division techniques. The general nature of meiosis (in all main groups of eukaryotes) means that, first, it was in the LECA. Second, we cannot exclude that it may be ancestral for all living cells (by the stage of LUCA). Cavalier-Smith (2010) suggests that mitosis and meiosis may have evolved together, and this could fit in well with the previous section about the origin of proteins. Similarly, Lehman (2003) argues for the 'extreme antiquity' of recombination, right back to the origins of life. It has been known for many years (e.g. Muller 1964) that without any form of recombination, organisms can accumulate slightly deleterious mutations and also that recombination (as during sexual reproduction) allows new recombinants, including some without any semi-deleterious mutations. This effect is also potentially reduced by large population sizes, such as may occur with bacterial cells, so the larger population sizes of akaryotes may mean that they have lost recombination.

It is an interesting question of how sexual reproduction arose—it cannot be for the long-term benefit of the lineage. We do need to be careful here; nearly all eukaryotes have meiosis and recombination, but separate sexes/genders arose much later—that is a very different issue. There is an aspect of 'cost' to meiosis in that many double-stranded breaks are made in the DNA, many of which do not lead to recombination (Bolcun-Filas et al. 2014). Rather, they are 'repaired' by an error-prone mechanism (so this can potentially introduce errors), although the repair of these double-strand breaks might be suspended during mitosis to prevent telomere fusion (Orthwein et al. 2014). There is the likely possibility that the earliest cells had some form of recombination (e.g.

see Gilbert 1978) that could have sped up evolution. Akaryotes have an apparent advantage in that they appear to be able to 'take up' existing genes by lateral gene transfer from other organisms, and this increases the advantages of meiosis in that way. (This means that there must be some reconsideration of the biological species concept for akaryotes.) Anyway, the main point here is that some process akin to recombination and meiosis may have been essential for early life; we do not know yet.

Translation in bacteria is virtually simultaneous with transcription; the mRNA may still be transcribed even when the mRNA is being translated to a protein. The timing for eukaryotes was initially measured using tritiated thymidine incorporated into mRNA and then observing how long before the mRNA appeared in the cytoplasm; the process of transfer to the cytoplasm (with its ribosomes) could take up to two hours, though it was faster (about thirty minutes) in some small yeast cells. Is there any immediate advantage of having this long-time period? As mentioned earlier, there does appear to be a reduction in intron number for eukaryotes with a short life cycle (Jeffares et al. 2006), and this delay might be related to the time required for processing the removal of each intron. So the question remains, is there a real advantage in going from a more or less immediate copying of mRNA, as in bacteria, to the long delay in eukaryotes? Or another way of thinking about this; is there no disadvantage in the long delay for eukaryotes (given a slower reproductive time of earlier cells)?

Epigenetics

There is now good evidence for several aspects of epigenetics in all the main subgroups of eukaryotes, but we do need some additional information to increase our confidence here. However, we won't repeat that discussion here, though it includes methylation, acetylation, and phosphorylation. DNA methylation is known in diverse eukaryotes from mammals to nematodes to ciliates, where methylation and hydroxymethylation are involved in DNA elimination.

There is also a strong link between epigenetic mechanisms and small RNAs (reviewed in Collins et al. 2010). Such RNA-directed modifications occur in other eukaryotes although the mechanisms may be less understood. In *Plasmodium*, long ncRNAs are involved in a mechanism similar to Xist. So our current conclusion is that epigenetics was also in the LECA (FRED), though more information would certainly be helpful here. So we see epigenetic mechanisms reported throughout eukaryotes. However, with our knowledge still growing in protist systems, it is still too early to say whether all epigenetic mechanisms stem from an ancestral one; but at the moment, it appears as if most of the epigenetic mechanisms occurred at least in LECA, if not in LUCA itself.

Timing Required for mRNA Processing (the Complexity of LECA)

The main point here is that under the usual model, it would take a very long time to get a complex LECA (FRED) from a prokaryote, and this is illustrated in Figure 7.5A. The high complexity of LECA means that if eukaryotes did arise by a fusion of a bacterium and an archaeon, then there must have been a very long time between mitochondrial uptake and the LECA. There might have to have been hundreds of millions of years for all the cytological and biochemical features in Box 7.1 to have evolved, and there is no obvious immediate selective advantage for many of them to have evolved later. There are several critiques of the fusion hypotheses, including Forterre (2013).

Alternatively, if a 'protoeukaryote' (an organism with nuclear/cytoplasmic division, multiple centers of replication, etc., as outlined in Box 7.1) took up the bacterial mitochondrial precursor, then this could have happened anywhere from relatively soon before LECA to a long time before. From any Bayesian perspective, this option (protoeukaryote origin) is more likely because there could be a relatively short time between the uptake of the endosymbiont to an intermediate time (option B in Figure 7.5) to a very long time (option C, Figure 7.5). This second alternative (a protoeukaryote origin) is therefore much more likely because the priors are very broad, which is not the case in the first alternative (fusion)—fusion would involve a very long time period (as mentioned, probably many hundreds of millions of years) to develop all the features of the LECA (see Box 7.1), none of which are explained directly by the fusion. At present, advocates of the fusion hypothesis do not appear to explain the many cytological features of the eukaryotic cell. We do need good hypotheses here.

Models for the Origin of Akaryotes from Eukaryotes

There are at least four proposals by which akaryotes may have arisen from a eukaryote ancestor: (1) thermoadaptation of bacteria, (2) via planctomycetes, (3) via virus evolution, or (4) the adaptation to higher temperatures (in archaea, as mentioned earlier). So first, there is the well-established 'thermoadaptation' model where many of the akaryote features (Forterre 1995) are considered responses to higher temperatures, where RNA is not that stable (Toffano-Nioche 2013). A second alternative is the progressive adoption of an akaryote lifestyle, with some groups (such as planctomycetes, deeply diverging bacteria) still showing signs of a nucleus, including a nuclear membrane and nuclear pores. This has been discussed by Fuerst (2013).

A third model (Figure 7.6) is based on the large viruses that currently inhabit giant amoeboid cells (Philippe et al. 2013), so such large viruses do occur. The model requires that the virus (presumably with a single center of replication) keep acquiring genes from already processed mRNA. If it were a cytoplasmic virus, it is likely to have lacked exon/intron processing capacity and was already able to reproduce in the cytoplasm (of a 'eukaryote-like' organism). Thus it would not really use/need the nuclear compartment; this is likely to be a significant advantage for some types of cells that did not take up other cells by phagocytosis. This model is discussed by Penny et al. (2014). The fourth model has been given earlier and is the adaptation to higher temperatures in archaea, possibly due to membranes functioning at higher temperatures.

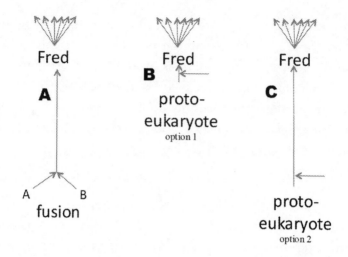

Fig. 7.5. Alternatives for the origin of the eukaryote cell, each with an endosymbiotic bacterium becoming the mitochondrion. The first hypothesis **(A)** is a fusion between an archaeal cell and a bacterial cell. Because of the high complexity of LECA (FRED), there would have to be a very long time between any such fusion and LECA in order for the many eukaryote features (including a nucleus, nuclear pores, spliceosomes, etc., to arise (see Box 7.1)). An alternative hypothesis, shown in **(B)** and **(C)**, is that there was a pre-existing 'protoeukaryote' cell (without a mitochondrion) that already had most of the features of eukaryotic cells and that this proto-eukaryotic cell took up the bacterium to form the endosymbiotic mitochondrion. From any Bayesian perspective, this option (B) appears more likely.

The first three models are well established and emphasize that there are good hypotheses about a eukaryote-to-akaryote transition. We make no value judgment to support any of them over the others, but we do not yet have an equivalent good model for the reverse transition (akaryote to eukaryote) that would give a positive (and immediate) selective advantage to all the intermediate stages illustrated in Box 7.1. But it is still the favorite hypothesis, akaryotes to eukaryotes.

It seems that the genes/proteins that are transferred to the nucleus (from mitochondria or plastids) do behave as 'eukaryote genes' in that they get eventually an exon/intron structure if transferred to the nucleus. So this implies that the exon/intron structure is normal/usual for eukaryotes—there appears to be no real advantage to the cell over which structure is adopted. However, this effect is perhaps limited to eukaryote cells with a longer life cycle. Jeffares et al. (2006) report that there is a loss of introns for eukaryotes with a short life cycle possibly because it may take two to three minutes to process/ remove each intron, and having many introns in a gene would be a significant delay for eukaryotes with a relatively short life cycle. This in itself is a very interesting point in that it would imply that there must have been eukaryotes with a reasonably long life cycle from the very earliest stage of eukaryotes (see De Nooijer et al. 2009). It could *not* have been that all the earliest 'eukaryotes' had a very short life cycle, or else there would be selection against introns. So there does appear to be selection against introns at the fastest eukaryote life cycles, and therefore this would not work as a general explanation for their origin.

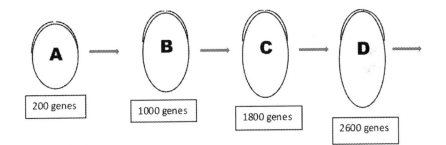

Fig. 7.6. One model for the origin of akaryotes from eukaryotes. A circular virus that inhabits the cytoplasm of an early protoeukaryote will have a certain number of genes. It successively accumulates more and more genes as copies until it has sufficient genes to be an independent cell. By being a cytoplasmic virus, it already has the machinery for replication in just one cellular compartment, and so it does not need the nucleus.

In principle, there is no reason why the same explanation holds for both bacteria and archaea when we consider that there are at least four models; one might apply for bacteria and another for archaea! It is interesting that there are two groups of akaryotes (bacteria and archaea) and only one group of eukaryotes, but this does not indicate the ancestral condition; it depends on the properties of the 'outgroup' (Figure 7.1), and this is still unknown. Perhaps the uptake of the mitochondrion was vital for modern eukaryotes. It would have given them an important advantage.

This 'model' does assume that bacteria, for example, with their much larger population sizes, are the agents for much innovation in biochemistry (this is a necessary consequence of Michael Lynch's model (Lynch and Abegg 2010)). In other words, they can have many more pairs of mutations that are neutral of themselves, but positive when two or three or more are combined. It is not surprising then that these akaryotes are able to carry out oxygen-dependent ATP formation (that eventually became mitochondria). They also are the beasties that carry out most photosynthesis, including through chloroplasts. They are brilliant little organisms.

Models for the Origin of Eukaryotes from Akaryotes

Perhaps this is the most popular current model among evolutionists. It is often assumed that there was a fusion of an archaeon and a bacterium, with the bacterium becoming the mitochondrion and the archaeon, in some unknown way, developing a complex nucleus—and all the other features of a eukaryote cell, including multiple centers of DNA replication! But this 'fusion' model does not explain any of the characteristic features of the eukaryotic cell (the list in Box 7.1). How do we explain how two compartments arise—mRNA capping, multiple centers of replication, exon/intron structure, etc.?

There are many other general properties of eukaryotes and akaryotes, but most of the others fit within the ideas discussed here. Perhaps some that do not fit directly are the origin of the very elegant operon structure in many bacteria, possibly as a response to higher temperatures with proteins having a limited life span (Glansdorf 1999). This would fit with the thermophilic origin of akaryotes, though it is not dependent on thermal adaptation. Pass!

Where Now?

So where does this leave us? In an important sense, none of us know the ancestral relationship between eukaryotes, bacteria, and archaea. Nevertheless, this has not stopped speculation about such a fundamental topic. To oversimplify perhaps, there may be a spectrum of ideas developing; at one end are those interested more in the origin of life and who are very open-minded about the relationships of akaryotes and eukaryotes. At the other end, there are the researchers mostly interested in the relationships of the three groups (and which we call the 'phylogeneticists'), and they appear to 'know' that akaryotes are ancestral. This is probably too simplistic, but all we aim to do here is to raise questions about the directions of the relationships.

Bacteria, in particular, appear to be extremely well adapted to their environments, and they also appear (given their very large population sizes) to be excellent at finding new metabolic pathways (such as photosynthesis and the use of oxygen in oxidative phosphorylation). Population sizes may effectively be very large in many bacteria, and there certainly is some specialization of function between the three main groups—even smaller eukaryotes appear to show a relatively high rate of bacterial consumption (Zubkov and Tarran 2008). The smaller cell size of akaryotes is also an advantage in many instances; they have better surface area to volume ratio (De Nooijer et al. 2009). Perhaps there is not yet the same appreciation of the advantages of the archaean lifestyle, apart from an appreciation of their adaptability and survival in minimal growth habitats.

Eukaryotes may not appear to have the same elegance as bacterial cells (lacking operons, for example), but eukaryotes have been effective in their own way. Their very large genomes have led to novel control mechanisms—often using RNA as part of their system. They can develop considerable complexity and have developed coenocytic cells (with a large number of nuclei per cell), large cells with a single large nucleus (for example, ciliates and the alga *Acetabularia*), and 'multicellular' forms. We define this latter class as having a large number of small cells, each cell with a single nucleus. This group (including plants, animals, and brown algae) has been very successful in allowing specialization of cells for different functions. Eukaryotes have a surprising dependence on ribonucleoproteins (RNPs). Perhaps there are important advantages in having a larger genome size in that it allows some novel (regulatory) genes to arise, some of which are potentially important for cell specialization. Complex multicellularity may have only developed in eukaryotes about 700 million years ago. Simple 'multicellularity' has arisen at least three times among bacteria, so some cell specialization is possible in akaryotes even if it is much more likely in eukaryotes. But simultaneously we do need good evolutionary mechanisms that give an advantage to all the features of the eukaryotic cell (as shown in Box 7.1). Some forms of epigenetics may provide such a mechanism.

There are certainly many experiments (including thought experiments) that offer a way forward. We would like to see experiments/measurements on a wide range of questions, including the time of transport of pre-mRNA through the various stages of processing in the nucleus. We would like to know more about the circular chromosome/linear chromosome alternatives that occur in bacteria. We would like to see more information on what might be rates of timing of DNA synthesis, rates of cell division, etc., in the LUCA. There are a great many questions for further study, but all do require keeping an open mind to the various options.

Whatever happens in the future, we must be open to the possibilities of the relationships between eukaryotes and akaryotes. There has perhaps been too much reliance on evolutionary trees even though we know that there are major difficulties at this deeper level. (Damn those mathematicians, spoiling biologists' fun in the evolutionary trees (e.g. Mossel and Steel 2004)!) However, any model proposed for their interrelationships must consider all the evidence and should be fully in accord with known evolutionary mechanisms (or formally present new ones). In particular, any mechanism must show short-term benefits for the possessor of the feature, and this latter aspect does present a problem with many models that derive eukaryotes from akaryotes. For example, what is the immediate short-term advantage of going back to multiple centers of replication? And having telomeres? A 'double fusion' might work—a fusion to give the nucleus/cytoplasm division and a second of this protoeukaryote with a bacterium to form the mitochondrion. I am not aware of anyone who has proposed such a double fusion, but maybe they have. Or maybe eukaryotes did come first—before the brilliant bacteria and archaea. We need hypotheses that lead to quantitative tests—and as we tell our first-year students, 'belief is the curse of the thinking class'. We must keep an open mind on the issues.

CHAPTER 8

What We Don't (Yet) Know about Evolution

I suppose the best answer to this Popperian question is 'quite a lot really'. So perhaps this is too general a question. Yes, we know that evolution has occurred (and is inevitable), but there is a lot to discover in the future, and perhaps it is best expressed as that we want to know what any given mutation might do in practice. We want to know the most desirable mutations and mutation combinations (and whether we are likely to get that combination). Anyway, it is an optimistic future because we have so much more to learn; and with the power of modern computers, we are going to go ahead much faster. In any Popperian system, we are going to have an exciting time, but biologists may have to learn to calculate effective and accurate tertiary structures and to actually predict the effect of any mutation!

We do need to be a bit careful here—everything depends on our genetics, but nothing depends on our genetics! Okay, we will consider the latter aspect first. As humans, we know that we are very flexible and that the phrase 'genetic determinism' does not describe our abilities well at all. But simultaneously we certainly do not expect a chimpanzee (or, for that matter, a fungus, a plant, a protist, a bacterium) to show full human-type abilities. So although we do reject 'genetic determinism', we also expect genetics to be vitally important. The following sections should be read in light of these two contrasting comments.

The main point of this chapter is that we have already learned a huge amount during the twentieth and early twenty-first centuries—and that it all fits in with 'causes now in operation'. The Lyell/ Darwin idea of explaining the past by causes that operate in the present is still operating; it appears to be the right way to analyze data for biological processes. The physical and chemical laws appear to be the same ones operating both in the past, in the present, and presumably into the future—these ideas are what we have to work with. So we should now continue with our Popperian travels and look forward to ideas that appear to be may be as yet unknown, but that are all quite reasonable based on our current knowledge.

Genetics of All Mutations

What we would like is to have more knowledge of the power of genetics but applied to whatever a particular mutant might do or how it might behave (including having no effect). Genetics has given us a simple one-dimensional (linear) structure, which we know and love. Reality (and the principles of chemistry) then translates this into a three-dimension tertiary structure, but which we do not (yet) fully know, though we are learning quite fast! As we said earlier in Chapter 2, it was a problem in the eighteenth century in that they had to assume that an *interior milieu* had to accumulate in the gonad tissue in order to pass on the tertiary structure of the heart, the kidneys, etc. We have made good progress in understanding tertiary structure of proteins, initially by X-ray crystallography but more recently by a variety of newer techniques as well, such as various cryo-techniques at very low temperatures. Yes, we do not yet know all the principles behind tertiary structure, but it is brilliant that we can use a lot of information and knowledge to reconstruct three-dimensional structures, and our computers are getting much better at predicting the structures from first principles (and from known structures). We still have quite a lot to learn here, but the subject area is moving quite fast, and our ability to predict and understand tertiary structures is improving all the time. We are a bit slower at understanding how tertiary structure changes with time (quaternary structure), but this knowledge is also improving with time.

Perhaps what we really want to know is to predict how any given mutation might behave—either in a sequence coding for proteins or in RNA that regulates the level of expression of genes. Perhaps we should give a classification of mutations first—there are (as we have said) mutations in the sequence of mRNA, in short RNAs (sRNA), and in 'long non-coding' RNAs (lincRNAs). Let's take just the former first, a mutation in an lncRNAs; and if it changes the amino acid being coded for, then we have a mutation in the sequence, which might (on average) be neutral, advantageous, or deleterious. There is some good progress here in that mutations that might be involved in catalytic sites can usually be identified. But we need to know (predict) what any given mutation might do. A number of mutations do not change the amino acid being coded for but might change the tRNA that places the amino acid, and some of these changes can be effective, especially in the regulation of protein levels. There are also combinations of mutations (for example in cancer cells), and so it is necessary to know what combinations of mutations might do as well. But first, we need to know what any mutation might do. Similarly, an apparently neutral change might affect binding by a sRNA. Regardless of this, will a specific mutation enhance a catalytic reaction, not affect the rate, be less effective on (but still carry out the reaction), or be inhibitory in at least some cells? For humans, at least perhaps the most likely area for this to occur is in the understanding of molecular medicine. There is a huge industry here, considering mutations in genes—we must eventually get to the predictive stage in this area. In addition, many show alternative splicing—it is estimated that over 90% of genes in humans can show this. What is the effect of alternative splicing?

So we still have quite a long way to go to predicting the full effects of a given mutation. Perhaps the first step is deciding whether or not a cofactor (and which cofactor) should be available to predict the proteins folding. Some proteins fold with a specific cofactor; others fold without them. Usually the protein has its three-dimensional structure determined with a cofactor already in place, and what the normal cofactor for a particular enzyme is usually known; that is part of knowing about the enzyme, but many proteins (such as ribosomal proteins) lack cofactors. Sometimes a specific protein will have a mutation, and it is known to be inhibitory, but that is not the same as predicting it will be inhibitory. Increasingly chemists are studying larger and larger molecules, and so it may be just a matter of time. But this still does not yet tell us what a particular mutation might do, and that is important to know for future studies. About 30% of human proteins appear not to be folded in a precise way, and perhaps it is best to treat them as 'disordered' proteins. Many of these interact with other proteins so may (in practice) not be 'disordered', but we still have a long way to go in understanding such proteins. Some proteins form dimers, trimers, etc. But there is a lot to be learned here (and that is positive news for the future). And (as we have already mentioned) we need to know much more about the interactions of proteins—which ones interact in practice.

It is not enough to know each step in a reaction—we need to know the pathway as well. So there is the question of predicting which step they catalyze. Chemists are usually very important at this point; they are used to predicting the rates of reaction of smaller compounds. But proteins are much larger to work with, so we will need to have even greater interaction between chemists and biochemists for this approach to work effectively. Perhaps the genetics was the easy part; once we knew the linear sequence of a gene (and it is readily available by sequencing genomes), it was easily possible to infer whether or not a change altered the sequence. But there are still too many 'unclassified' proteins whose functions we have little idea of, so we are still learning! Now we need more information about the effective tertiary structure of the proteins and the variations in structure that occur in practice, though that is still not enough. But we must be able to predict the tertiary structure of proteins, and some very good progress is being made on this issue. Some good progress is also being made by determining what proteins are essential when cells are supplied all the nutrients that they need, such studies on cells (Hutchinson et al. 2016) are essential.

Multicellularity

A related question is one of the most important issues and is the mechanism by which different genes are expressed at different levels in the different cells of multicellular organisms. It does *not* appear that the simple linear structures of classical genetics will explain the problem of why different cells in multicellular plants and animals express different proteins. As we have said, there is a real question as to why most root cells do not express photosynthetic proteins, but leaf cells do. Why? And the genes are there (many root cells can give rise to leaf cells). We would all like to have a simple answer, as simple as basic genetics with its linear description of mutations. But it appears that the mechanism is dependent on both the three-dimensional structure of proteins and on small and long RNA molecules, and that will always complicate things—(unfortunately) tertiary structure is much more complex than linear structure. But this is an area ripe for progress in understanding.

Other genes (for example, hemoglobins in vertebrates) may be expressed at different stages of development: embryo, fetal, and adult. Could we manage to delay the onset of a deficient mutant adult hemoglobin? A hemoglobin expressed in an adult cell might be a deleterious mutant, but the fetal one might be fine. Should we aim to keep expressing the earlier one? This may be another way of understanding some of the questions about why different genes are expressed at different stages of development. We could also improve identification of species considerably by sequencing genes at a room temperature reaction; this would both identify species and aid identification.

We might say that some genes are 'turned off' (and not expressed), but that is basically an observation, not an explanation. Other genes could be regulated at different levels—maybe some of us are just luckier than others. As far as we know, all the cells in a multicellular organism have the same genes. There may be quite simple reasons why multicellular organisms have genetically identical (or at least genetically very similar) cells, but with very different functions; and it is possibly solvable more or less with current technologies. For most plant cells (at least), the cells can change between types; root cells can become stem and leaf cells, for example. For animal cells, we call some of them 'stem cells', and they are the cells that can become specialized for different functions—why or how? Other cells are more specialized. No doubt there had to be quite a long time period before cells learned to express some proteins at different stages of development, but not others. This is normal gradualism/continuity!

For example, multicellular plants and animals are a relatively late development, but animals arose before multicellular plants, so animals probably had learnt the mechanisms earlier than plants. There is no reason to assume that plants and animals use the same techniques to regulate gene expression levels, but perhaps they do. Or are there differences in mechanisms between plant and animal multicellularity? And what about fungi? Most multicellular organisms are eukaryotes. About the only bacterium that has apparently learned to express different genes in different cells are in the genus *Nostoc* (cyanobacteria, blue-green algae), which have most photosynthetic cells, but a few cells that 'fix' nitrogen instead of being photosynthetic (and are very specialized for fixing nitrogen). So most of the cell specialization is in the eukaryote cells, and most akaryote organisms only produce a single cell type, though the levels of proteins in these cells are variable with the growth conditions; and this must be controlled in some way. Perhaps the greater diversity of multinucleate eukaryote cells depends on their more complex genomes. But this is only a description, not a real answer. Thus eukaryotes have shown all sorts of cell specializations, and they have many short RNA (sRNA) molecules. Maybe the cells of eukaryotes, most of which have much more DNA, allow many more sRNA molecules. There is a lot to be learned here, and that is good news!

Different Levels of Proteins with Conditions

Having introduced the problem of different levels of proteins in different cells, we come to the more general problem (and almost certainly much older) of the regulation of protein levels/expression. Some good progress is being made here, with the smaller RNA molecules turning out to be very important. For convenience, we tend to divide the RNA into groups: the mRNAs, the lncRNAs (long non-coding), short RNAs (these have a cacophony of different abbreviations—we will just use 'sRNAs' for the shorter ones), and a few mega RNAs. However, we still do not have a good way of predicting how a specific variant RNA might behave and with the proteins it will interact with. It is probably helpful to divide mutations up into at least three groups. There are mutations

affecting specific catalytic reactions,
altering the general level of gene expression, and
altering specific gene levels.

Perhaps that at least would be a good start. And there is the specialized (very large) proteasome for breaking down proteins, especially misfolded ones. Proteins for destruction are marked with the protein ubiquitin and then broken down. But maybe it is more general than this; maybe RNA molecules are involved. Perhaps this is a question I have been leading up to here—we all want to know what a particular mutation might do! This includes the interactions between species—for example, the invasion of a pathogen; this can affect the level of expression of both host and pathogen proteins. Fortunately, we now have the ability to sequence the RNA molecules directly, and this is a good indication of the level of the RNAs, which can change quite dramatically between cells in different conditions, including sRNAs, mRNAs, and lincRNAs.

The question of the regulation of genes is the very general question that is still key to our understanding of how the cell really works. Again, we would like a good simple mechanism (as simple as classical genetics), but we do not (yet) seem to have a method that explains the different levels of gene expression simply. As we have said, it does seem that small RNAs (sRNAs) are important, though we lack a good method of predicting how variations in their tertiary structure can affect the levels of any particular gene. They can lead to rather different gene expressions by the destruction of some mRNAs—it is a combination of the small RNAs and some proteins (such as argonautes) that leads to the destruction of the mRNA. Does it occur (in eukaryotes) before or after splicing? But until we can predict the probable effect of a mutation, there is little that we really know! Soon perhaps. This is a crucial problem to understand, and until we can predict all the features of mutations, we are still (unfortunately) at a much lower level of understanding.

Systems Chemistry and Systems Biology

These are two relatively new techniques, and perhaps we will consider systems chemistry first. I could say that when I was an undergraduate chemistry major, we studied what I called 'clean' chemistry—the optimal amount of compound A and the optimal amount of compound B and the optimal conditions to make compound C. Nowadays chemists study what I call 'dirty' chemistry—what a mixture of chemicals would do. On no (say the chemists)—we call that systems chemistry; what reactions do you get in a mixture of chemicals? Okay, I agree, but I still like to call it 'dirty' chemistry, though I have to agree that this systems approach is far more interesting and relevant (though we did have to know the details of reactions first). So we need clean chemistry first and then the study of mixtures. This is good progress, but we need to go much further.

Similarly, there is a new study of biophysical science called 'systems biology', and that is very promising in this regard. It studies the behavior of systems of molecules, and this is a good start, and it includes the folding of macromolecules. But we need to go still further and predict or tell what a given system might do. Both these approaches (systems chemistry and systems biology) are very promising, and good progress is expected. We have a lot to look forward to here.

Brain Science

What mutations make us human? Okay (and as we have seen), centuries ago, René Descartes considered humans as distinct, and science didn't really apply to humans. But he also had the distinction that in some areas science was the criterion for how things worked, but not how the system arose (see Figure 1.2). As we have seen in Chapter 1, this initially gave science an area where it was responsible, but soon Pierre-Simon Laplace for physical sciences showed that science could explain the origin of solar systems. Later Charles Darwin did a similar step for biological sciences—and he advocated that science applied fully to humans. We have examined language ability as perhaps the most distinctive aspect of humanity and found that chimpanzees and gorillas (as the two closest relatives of humans) have very good sign language ability (but no 'voice box'). As such, we have suggested that 'normal' mutations would lead to humans (including their voice box).

But 'brain science'—we still cannot say how (or why) 'memories' occur, though it certainly evolved long before humans. Okay, we know that microelectrodes were first tried out in long plant cells (*Nitella*). This cell had an action potential that moved along the cell at about 1 centimeter every six to ten seconds. Once microelectrodes were better developed and functional in such *Nitella* cells, they were tried out on the long nerves of the giant squid, and they showed that they transmitted action potentials very much faster—and then history begins! From that time onwards, we have a good record of transmission studies in nerve cells! It is brilliant that animal cells take a feature of (all?) large cells (such as the plant *Nitella* cells) and modified them to give nerve cells, with their much faster transmission of stimuli. So this shows that cells took a 'normal' function and modified it—good science.

But this still does not tell us at all how memories are formed—just that transmission of stimuli is a natural phenomenon. What is the basis of our memories? Is it some physical structure that we might still eventually understand? Is the basis of learning some physical arrangement of cells that are formed? Or is it the connections between cells that are important? Or are we barking up the wrong tree, and it is something quite different? We do know that humans learn sequentially—very young babies learn to vocalize by steps. There is a lot to learn here fortunately.

The Origin of 'Species'

The species concept probably needs to be updated, and as we have seen, it is a pre-evolutionary concept and unfortunately could be considered quite static. We now think of it more dynamically and think of varieties, subspecies, species, subgenera, genera, etc., as part of a continuum. Perhaps the first issue is just normal hybridization, which is quite usual for at least plants. So we need a more dynamic view allowing hybridization. But we have to admit that animals generally do not have a 'species concept' and tend to 'breed' with whatever is available (Paré 1585). It would be good to know, at the molecular level, why many hybrids are sterile. The second issue is the acceptance that 'lateral gene transfer' is quite common, at least in bacteria. So maybe a 'network' is a better model for them (and for hybridization). This is a small step for 'tree thinking' to 'networks' and allows us to have a more dynamic view, but it is still 'causes now in operation'. One of the problems is that we still do not know how genes/proteins evolve over time; we certainly do not expect every tree of genes to be the same—mutations are random, and we expect the same mutation to occur on different lineages. Many genes are unique to a given species, so something is happening here.

A more rigid species concept may be loved by a few conservationists, but we have to allow some ability to evolve over time and allow some adaptation to climate change. Nevertheless, the fixed 'species concept' has been limiting to progress in some ways in that it appears to have inhibited the recognition of hybridization processes leading to new species formation, certainly in animals, fungi, and plants. The New Zealand Department of Conservation is reputed to have a policy of 'shoot the hybrids' even though hybrids may be part of the process of adaptation to climate change and to speciation! Really, the concept of species is a dynamic concept. We have seen that the 'species concept' is a relatively early invention during the seventeenth century, and it is certainly a very useful concept. So the species concept is very useful—but we should consider not just the current version as sacrosanct.

Protein/Enzyme Evolution

We need to study protein evolution more carefully. We tend to assume that each enzyme carries out just one task, but in reality catalysis (as studied by chemists) is not as specific as that. Thus it appears that proteins should not be fully specific for what they catalyze. Most enzymes are almost certainly not 100% specific, so it would be very interesting to know more about what reactions they possibly carry out (at least weakly).

Combined with this, mutation is random, and changes to the sequence occur when any change is not corrected. We need to study more the different possibilities and observe what a given enzyme might possibly do how it might evolve. We expect (predict?) that a given enzyme (especially when duplicated) may gradually evolve to a different function—does it? There is a lot to learn here! But we expect that some (proteins at least) do change their optimal reaction that they catalyze.

The Origin of Life

We are still not sure how life arose in the first place, though that has not stopped billions of dollars being spent in rocketry, ostensibly to study things like the potential for life on other planets. (Okay, that is an interesting question!) However, for the present we will just assume that if life really is a 'natural property of matter', we are really only interested here on earth-based life. Life may well have also arisen on other planets, and it will be very interesting to see whether such life is composed of the same pairing of nucleotides as here on earth and whether it has the same amino acid code as earth-based life. It is the latter aspect that we are particularly interested in. There are reported to be some 10^{27} possible genetic codes (even for the same twenty amino acids), so it is extremely important to study whether any extraterrestrial life had the same code as life on earth. I suspect not, but maybe I am a little prejudiced on this issue—I guess that I am 'earth-bound' (and see life as a natural property of matter (see Chapter 4); it would be nice to know). However, it is certainly expected that the origin of life is a process with very many intermediate stages, and it will be interesting (eventually) to determine how far along in evolution various groups on other planets are.

The first aspect is that it is generally considered that the first of the modern polymers was RNA (but there could have been previous polymers, though there is no evidence for this). There is good evidence for RNA being the first of the modern polymers—for example, ribosomes (which synthesize proteins by joining amino acids) are composed of both RNA and proteins. However, when the ribosome structure was determined, it was an 'RNA machine' (Yusupov et al. 2001)—all the catalytic sites were RNA (catalytic sites within about 8–9 Å were RNA (an Å is 10^{-10} of a meter); catalytic reactions are generally catalyzed within 1 Å). So we don't know whether RNA was the first polymer, but it does appear to come before proteins and DNA.

Now that the RNA world has such a good following, we must determine a pathway for synthesis of the components, and good progress has been made here. It was assumed initially that we made the nucleotides, then the sugar, and finally phosphate was added. However, it was shown (Attwater et al. 2013) that a possible (prebiotic) synthesis produced all the components and made the ribonucleotide phosphates directly. We also know that the ribose sugar (and deoxyribose) is important in that it allows G to pair with C and of A with either T (thymine in DNA) or U (uracil in RNA) (Eschenmoser 1999). So the sugars are important too; they are part of the structure.

An additional aspect is the origin of the triplet code, and here the Eigen limit (Chapter 4) is important, allowing longer and longer RNA models to be copied. It is assumed that copying doublets or triplets of RNA would increase the accuracy of copying. This still needs to be tested—that is, using a precursor of a triplet code should allow longer RNA molecules to be copied (Penny and Zhong 2015). Theoretically, this development of the code should allow longer and longer RNA molecules to be copied accurately. And then a second (duplicated) copy starts making short peptides, and then the proteins.

There is good progress in 'designing' a minimal cell (e.g. Hutchinson et al. 2016), and there seems to be some 'essential' proteins whose functions we still don't know. Rapid progress is expected here. It would be good if we could ascertain the temperature at which life arises (see earlier), but until we can do this, we really cannot tell the experimentalists what they should try.

One of the issues here is that traditionally it seems that life (or evolution) is seen as a 'selfish' program whose benefits go only to the possessor of advantageous mutations. However, in reality, evolution is much more cooperative than that. For example, we need to consider in which organisms each of the amino acids and vitamins are made. For example, humans require at least nine amino acids from their diet and thirteen vitamins (vitamins are usually cofactors that are provided by the diet). At least two cofactors are now currently made only by akaryotes, but plants have learned how to bypass the need for vitamin B_{12}; it is used by only three enzymes, and for example, land plants have learned to bypass the requirement (Croft et al. 2005).

Almost certainly there have always been 'cells' that could not make some amino acids—for example, tryptophan or phenylalanine (two essential amino acids for proteins). This is because some amino acids are 'leaky', and their use cannot be monopolized by any one organism. There is a positive interaction between many cells—and it is now accepted that humans have trillions of bacterial cells, mostly in their intestinal system, but also on their skin. In essence, this allows other cells to make some of the essential amino acids. In addition, there always seem to have been predators around (De Nooijer et al. 2009), so some of these cells will always have access to amino acids, or cofactors, synthesized by other organisms. This is sometimes known as the 'Black Queen' hypothesis, that some organisms will lose genes for synthesizing some compounds. However, I guess that I see it is even more general—we virtually cannot see life arising if it means every organism synthesizing everything it needs; life is much more cooperative than that. This message needs to be more widely appreciated; evolution is also cooperative. Unfortunately, there is no 'money to be made' for solving the origin of life question—so it is somewhat of an intellectual pursuit only. There does seem a tendency for humans to seem a little pecuniary here!

The Future

There is reputed to be a Chinese curse: 'may you live in interesting times'! Yes, we live in very interesting times, and there are so many things that we must learn—bring on the future! (To hell with the curse then.) We have made very good progress during the twentieth and early twenty-first centuries—we have kept learning new things and making new discoveries. However, we do not yet know whether the future of life is carbon based (like us) or whether it is more likely to be silicon based (like computers and robots). If computers learn to design and make other computers and if they also learn to make their own power sources, then that is a real question. But will the robots take over? If robots learn to make other robots, then maybe the future is 'silicon based' and not 'carbon-based' life forms. None of us know for sure—maybe we will become 'tired' of life. It was nice when we 'believed' that we were made by some unknown (unknowable) God(s) who ruled the universe with a deterministic passion. But that hope/desire has long gone for most people. But as far as we know, we still need humans.

But what we have learned is that all kinds of 'science' (including social sciences) fit in with 'causes now on operation'. However, an important issue is that theologians can no longer burn us at the stake if the results get the wrong answer. (Okay, some ISIS supporters might try the equivalent, but fortunately we know that they are really a small minority; most people are more rational than that, and they use 'evidence' on which to base at least some of their conclusions.) But as we have seen, 'belief is still powerful—even among so-called scientists.

We really can't 'stop evolution'; it just goes on and on—dammit. For example, for the influenza virus, we select new strains for immunization every six months—once for the northern hemisphere winter and then again for the southern hemisphere winter. Yes, the strains have changed over that time, and new varieties and recombinations have arisen. So we select the main strains for making antibodies and which we give people at the beginning of winter to give them immunity against the common strains of influenza viruses that occur at the time. Okay, RNA viruses have a higher mutation rate than DNA viruses, so they do evolve faster; they can't compare the sequence of the old strand against the sequence of the new strand. (DNA-based organisms can always tell the sequence of the newly synthesized strand from the older strand). But evolution goes on and on and on.

So yes, we need to predict whatever a given mutation might do. What are the alternatives? Maybe some new mutation will give us new abilities. We do not yet know. If this were true, then we should make that mutant and ensure that it spreads through the populations. I guess that I am an optimist. We will continue to learn and to cooperate. Science today is truly international, and there is only one reality. We have so much more to learn; the future is quite fantastic. Almost certainly we will learn to predict what new mutations will help us and our much-vaunted intelligence.

References

Allentoft M.E. et al. (2014), 'Extinct New Zealand Megafauna Were Not in Decline before Human Colonization', *Proc. Natl Acad. Sci. USA*, 111: 4922–4927.

Alvarez, L. W. et al. (1980), 'Extraterrestrial Cause for the Cretaceous-Tertiary Extinction—Experimental Results and Theoretical Interpretation', *Science*, 208: 1095–1108.

Andersen, J. C. (1969), *Myths and Legends of the Polynesians* (Tokyo).

Anderson, A. et al. (2010), 'Faunal Extinction and Human Habitation in New Caledonia: Initial Results and Implications of New Research at the Pindai Caves', *J. Pacif. Arch.*, 1: 89–109.

Appel, T. A. (1987), *The Cuvier-Geoffroy Debate: French Biology in the Decades before Darwin* (Oxford).

Archibald, J.D. et al. (2011), 'Protungulatum, Confirmed Cretaceous Occurrence of an Otherwise Paleogene Eutherian (Placental?) Mammal', *J. Mamm. Evol.*, 18: 153–161.

——, et al. (2010), 'Cretaceous Extinctions: Multiple Causes', *Science*, 328: 973–973.

Attwater, J. et al. (2013), 'In-Ice Evolution of RNA Polymerase Ribozyme Activity', *Nature Chem.*, 5: 1011–1018.

Bolcun-Filas, E. et al. (2014), 'Reversal of Female Infertility by Chk2 Ablation Reveals the Oocyte DNA Damage Checkpoint Pathway', *Science*, 343: 533–536.

Bowler, P. J. (1986), *Theories of Human Evolution: A Century of Debate, 1844–1944* (Baltimore).

—— (2003), *Evolution: The History of an Idea*, 3rd edn. (Berkeley).

Boussau, B. et al. (2008), 'Parallel Adaptations to High Temperatures in the Archaean Eon', *Nature*, 456: 942–945.

Briggs, A. W. et al. (2009), 'Targeted Retrieval and Analysis of Five Neanderthal mtDNA Genomes', *Science*, 325: 318–321.

Burkhardt, R. W., Jr. (1977), *The Spirit of System: Lamarck and Evolutionary Biology* (Cambridge).

Burkhardt, F., and Smith, S. (1986), *The Correspondence of Charles Darwin: Vol. 2* (Cambridge).

Cann, R. L. et al. (1987), 'Mitochondrial DNA and Human Evolution', *Nature*, 325: 31–36.

Cavalier-Smith, T. (2010), 'Origin of the Cell Nucleus, Mitosis and Sex: Roles of Intracellular Coevolution', *Biol. Dir.*, 5: 7.

Chambers, C. (1844), *Vestige of the Natural History of Creation*, repr. in 1994 (Chicago).

Claramunt, S., and Cracraft, J. (2015), 'A New Time Tree Reveals Earth History's Imprint on the Evolution of Modern Birds', *Sci. Adv.*, 1:e1501005.

Cobb, M. (2000), 'Reading and Writing The Book of Nature: Jan Swammerdam (1637–1680)', *Endeavour*, 24: 122–128.

Codron, D. et al. (2012), 'Ontogenetic Niche Shifts in Dinosaurs Influenced Size, Diversity and Extinction in Terrestrial Vertebrates', *Biol. Lett.*, 8, 620–623.

Collins, L. J. et al. (2010), 'The Epigenetics of Non-coding RNA', in T. Tollefsbol (ed.), *Handbook of Epigenetics* (Oxford): 49–61.

——, and Penny, D. (2005), 'Complex Spliceosomal Organization Ancestral to Extant Eukaryotes', *Mol. Biol. Evol.*, 22: 1053–1066.

Comte, A. (1835), *Cours de Philosophie Positive*, tome II (Paris), reviewed in *Edinburgh Review* (1838): 271–308.

Croft, M. T. et al. (2005) Algae Acquire Vitamin B-12 through a Symbiotic Relationship with Bacteria. *Nature*, 438: 90–93 <DOI: 10.1038/nature04056>.

Cuvier, G. (1817), *Essay on the Theory of the Earth*, 3rd edn., trans. Prof. Jameson, repr. in 1978 (New York).

Daly, T. et al. (2013), 'In Silico Resurrection of the Major Vault Protein Suggests It Is Ancestral in Modern Eukaryotes', *Gen. Biol. Evol.*, 5: 1567–1583.

Darwin, C. (1969), *The Autobiography of Charles Darwin 1809–1882* (with original omissions restored), N. Barlow (ed.) (New York).

—— (1859), *On the Origin of Species* (London).

—— (1845), *The Voyage of the Beagle*, 2nd edn., in *Harvard Classics*, 29 (1909).

Darwin, E. (1791), *The Botanic Garden*, repr. in 1973 (London).

Dean, D. R. (1992), *James Hutton and the History of Geology* (Ithaca).

De Beer, G. (ed.) (1960), *Darwin's Notebooks on the Transmutation of Species*, *Brit. Mus. (Nat. Hist.) Hist. Ser. 2*, no. 2: 27–73.

De Nooijer, S. et al. (2009), Eukaryotic Origins: There Was No Garden of Eden? *PLoS One, 4:* e5507.

Descartes, R. (1637), *A Discourse on Method*, trans. J. Veitch, repr. in 1957 (London).

Desmond, A, and Moore, J. (2010), *Darwin's Sacred Cause: Race, Slavery and the Quest for Human Origins* (London).

Dewar, R. E. et al. (2013), 'Stone Tools and Foraging in Northern Madagascar Challenge Holocene Extinction Models', *Proc. Natl. Acad. Sci. USA*, 110: 12583–12588.

Diekmann, Y., and Pereira-Leal, J. B. (2013), 'Evolution of Intracellular Compartmentalization', *Biochem. J.*, 449: 319–331.

Duncan, R. P., Boyer, A. G., and Blackburn, T. M. (2013), 'Magnitude and Variation of Prehistoric Bird Extinctions in the Pacific', *Proc. Natl Acad. Sci. USA*, 110: 6436–6441.

Eigen, M., and Schuster, P. (1978), 'The Hypercycle: A Principle of Natural Self-Organization, Part C the Realistic Hypercycle', *Naturwissenschaften*, 65: 341–369.

Egel, R., and Penny, D. (2007), On the Origin of Meiosis in Eukaryotic Evolution, in R. Egel and D. H. Lankenau (eds.), *Recombination and Meiosis: Models, Means and Evolution* (Berlin) 249–288.

Eldredge, N., and Gould, S. J. (1972), 'Punctuated Equilibria: An Alternative to Phyletic Gradualism', in T. J. M. Schopf (ed.), *Models in Paleobiology* (San Francisco), 82–115.

Embley, T. M., and Martin W. (2006), 'Eukaryotic Evolution, Changes and Challenges', *Nature*, 440: 623–630.

Erhardt, H., and Knop, D. (2005), *Corals: Indo-Pacific Field Guide* (Frankfurt).

Eschenmoser, A. (1999), 'Chemical Etiology of Nucleic Acid Structure', *Science*, 284: 2118–2124.

Farley, J. (1977), *The Spontaneous Generation Controversy from Descartes to Oparin* (Baltimore).

Ferris, S. D. et al. (1981), 'Extensive Polymorphism in the Mitochondrial DNA of Apes', *Proc. Natl Acad. Sci. USA*, 78: 6319–6323.

Firestone, R. B. et al. (2007), 'Evidence for an Extraterrestrial Impact 12,900 Years Ago That Contributed to the Megafaunal Extinctions and the Younger Dryas Cooling', *Proc. Natl Acad. Sci. USA*, 104: 16016–16021.

Forterre, P. (2013), 'The Common Ancestor of Archaea and Eukarya Was Not an Archaeon', *Archaea*, 2013: 372396.

—— (1995), 'Thermoreduction: A Hypothesis for the Origin of Prokaryotes', *CR Acad. Sci. Paris*, 318: 415–422.

Fuerst, J. (2013), 'The PVC Superphylum: Exceptions to the Bacterial Definition?' *Antonie Van Leeuwenhoek*, 104: 451–466.

Gilbert, W. (1978), 'Why Genes in Pieces?' *Nature*, 271: 501.

Glansdorf, N. (1999), 'On the Origin of Operons and Their Possible Role in Evolution towards Thermophily', *J. Mol. Evol.*, 49: 432–438.

Gould, S. J. (1977), *Ontogeny and Phylogeny* (Cambridge).

Grady, J. M. et al. (2014), 'Evidence for Mesothermy in Dinosaurs', *Science*, 344: 1268–1272.

Green, R. E. et al. (2010), 'A Draft Sequence of the Neanderthal Genome', *Science*, 328: 710–722.

Greene, J. C. (1959), *The Death of Adam: Evolution and Its Impact on Western Thought* (Iowa).

Hawgood, B. J. (2003), 'Francesco Redi (1626–1697): Tuscan Philosopher, Physician and Poet', *J. Medical Biog.*, 11: 28–34.

Harvey, W. (1651), *Exercitationes de generatione animalium* (The Generation (Reproduction) of Animals), trans. the University of Toronto.

Herbert, S. (2005), *Charles Darwin, Geologist* (Ithaca).

Hildebrand, A. R., Penfield, G. T., et al. (1991), 'Chicxulub Crater—A Possible Cretaceous Tertiary Boundary Impact Crater on the Yucatan Peninsula, Mexico', *Geology*, 19: 867–871.

Hone, D. W. E, and Benton, M. J. (2005), 'The Evolution of Large Size: How Does Cope's Rule Work?' *Trends Ecol. Evol.*, 20: 4–6.

Hordijk, W., and Steel, M. (2013), 'A Formal Model of Autocatalytic Sets Emerging in an RNA Replicator System', *J. Systems Chem.*, 4: 3.

Hu, Y. et al. (2005), 'Large Mesozoic Mammals Fed on Young Dinosaurs', *Nature*, 433: 149–152.

Hurles, M. E. et al. (2003), 'Untangling Pacific Settlement: The Edge of the Knowable', *Trends Ecol. Evol.*, 18: 531–538.

Hutchison, C. A., III et al. (2016), 'Design and Synthesis of a Minimal Bacterial Genome', *Science*, 351 <DOI: 10.1126/science.aad6253>.

Hutton, J. (1795), *Theory of the Earth: With Proofs and Illustrations*, 2 vols. (London).

Huxley, J. (1942), *Evolution: The Modern Synthesis* (London).

Isley, D. (1994), 'Helmont', in *One Hundred and One Botanists* (Iowa).

Jacob, F. (1973), *The Logic of Life: A History of Heredity* (New York).

——, and Monod, J. (1961), 'Genetic Regulatory Mechanisms in the Synthesis of Proteins', *J. Mol. Biol.*, 3: 318–356.

Javaux, E. J. et al. (2010), 'Organic-Walled Microfossils in 3.2-Billion-Year-Old Shallow-Marine Siliciclastic Deposits', *Nature*, 463: 934–938.

Jeffares, D. C. et al. (2006), 'The Biology of Intron Gain and Loss', *Trends Genet.*, 22: 16–22.

—— et al. (1998), 'Relics from the RNA World', *J. Mol. Evol.*, 46: 18–36.

Johnson, C. (2009), 'Megafaunal Decline and Fall', *Science*, 326: 1072–1073.

Ke, Y. H. et al. (2001), 'African Origin of Modern Humans in East Asia: A Tale of 12,000 Y Chromosomes', *Science*, 292: 1151–1153.

King, J. L., and Jukes, T. H. (1969), 'Non-Darwinian Evolution', *Science*, 164: 788–798.

King-Hele, D. (1999), *Erasmus Darwin: A Life of Unequalled Achievement* (London).

Köhler, W. (1973), *The Mentality of Apes*, trans. E. Winter (London).

Lamarck, J. B. (1809), *Zoological Philosophy*, trans. H. Elliott (Chicago, 1984).

Lane, N. and Martin, W.F. (2012), The origin of membrane bioenergetics. Cell 151:1406–1416.

Lehman, N. (2003), 'A Case for the Extreme Antiquity of Recombination', *J. Mol. Evol.*, 56: 770–777.

Lewontin, R. C. (1974), *The Genetic Basis of Evolutionary Change* (New York).

Lloyd, G. T. et al. (2008), 'Dinosaurs and the Cretaceous Terrestrial Revolution', *Proc. R. Soc. B.*, 275: 2483–2490.

Lockhart, P. J. et al. (2014), 'We Are Still Learning about the Nature of Species and Their Evolutionary Relationships', *Annals Missouri Botanical Garden*, 100: 6–13.

Lockley, M. G., and Rainforth, E. C. (2002), 'The Track Record of Mesozoic Birds and Pterosaurs: An Ichnological and Paleoecological Perspective', in L. M. Chiappe and L. M. Witmer (eds.), *Mesozoic Birds: Above the Heads of Dinosaurs* (Berkeley), 405–418.

Longrich, N. R. et al. (2011), 'Mass Extinction of Birds at the Cretaceous–Paleogene (K–Pg) Boundary', *Proc. Natl Acad. Sci. USA*, 108: 15253–15257.

Lopez dos Santos, R. A. et al. (2013), 'Abrupt Vegetation Change after the Late Quaternary Megafaunal Extinction in Southeastern Australia', *Nature Geoscience*, 6: 627–631.

Lyell, C. (1830–1833), *Principles of Geology: Being an Enquiry How Far the Former Changes of the Earth's Surface Are Referable to Causes Now in Operation* (London), repr. in 1990 (Chicago).

Lynch, M, and Abegg, A. (2010), 'The Rate of Establishment of Complex Adaptations', *Mol. Biol. Evol.*, 27: 1404–1414.

Matthew, P. (1831), *On Naval Timber and Arboriculture* (London).

Mayr, E. (1991), *One Long Argument: Charles Darwin and the Genesis of Modern Evolutionary Thought* (Cambridge).

—— (1985), 'Darwin's Five Theories', in D. Kohn (ed.), *The Darwinian Heritage* (Princeton), 755–772.

——, and Provine, W. (eds.) (1980), *The Evolutionary Synthesis: Perspectives in the Unification of Biology* (Cambridge).

Mosimann, J. E., and Martin, P. S. (1975), 'Simulating Overkill by Paleoindians', *Amer. Sci.*, 63: 304–313.

Mossel, E., and Steel, M. (2004), 'A Phase Transition for a Random Cluster Model on Phylogenetic Trees', *Math. BioSci*, 187: 189–203.

Moulton, V. et al. (2000), 'RNA Folding Argues against a Hot-Start Origin of Life', *J. Mol. Evol.*, 51: 416–421.

Muller, H. J. (1964), 'The Relation of Recombination to Mutational Advance', *Mutatn. Res.*,1: 2–9.

Murphy, J. B. et al. (eds.) (2002), *Komodo Dragons: Biology and Conservation* (Washington, DC).

Murray, M. (2003), 'Overkill and Sustainable Use', *Science*, 299, 1851–1853.

O'Gorman, E. J., and Hone, D. W. E. (2012), 'Body Size Distribution of the Dinosaurs', *PLoS ONE*, 7: e51925.

Onoue, T. et al. (2012), 'Deep-Sea Record of Impact Apparently Unrelated to Mass Extinction in the Late Triassic', *Proc. Natl Acad. Sci. USA*, 109: 19134–19139.

Oparin, A. (1968), *Genesis and Evolutionary Development of Life* (New York).

Orthwein, A. et al. (2014), 'Mitosis Inhibits DNA Double-Strand Break Repair to Guard against Telomere Fusions', *Science*, 344: 189–193.

Ospovat, D. (1981), *The Development of Darwin's Theory: Natural History, Natural Theology, and Natural Selection, 1838–1859*. Cambridge.

Paley, W. (1802), *Natural Theology: Or, Evidences of the Existence and Attributes of the Deity* (London).

Paré, A. (1585), *Monsters and Marvels*, trans. and intro. J. L. Pallister (Chicago, 1982).

Patterson, F., and Linden, E. (1981), *The Education of Koko* (New York).

Penny, D. (2014), 'Cooperation and Selfishness Both Occur during Molecular Evolution', *Biology Direct.*, 10:26 <DOI: 10.1186/s13062-014-0026-5>.

—— (2009a), 'Charles Darwin as a Theoretical Biologist in the Mechanistic Tradition', *Trends in Evolutionary Biology*, 1: e1.

—— et al. (2003), 'Testing Fundamental Evolutionary Hypotheses', *J. Theoret. Biol.*, 223: 377–385.

——, and Zhong, B. (2015), 'Two Fundamental Questions about Protein Evolution', *Biochimie.*, 119: 278–283.

Philippe, P. et al. (2013), Pandoraviruses: Amoeba Viruses with Genomes Up to 2.5Mb Reaching That of Parasitic Eukaryotes, *Science*, 341: 281–286.

Pinker, S. (2011), *The Better Angels of Our Nature: Why Violence Has Declined* (New York).

Poczai, P. et al. (2014), 'Imre Festetics and the Sheep Breeders' Society of Moravia: Mendel's Forgotten "Research Network"', *PLoS Biol.*, 12: e1001772.

Pond, C. M. (1977), 'The Significance of Lactation in the Evolution of Mammals', *Evolution*, 31: 177–199.

Popper, K. R. (1945), *The Open Society and Its Enemies*, vol. 1 (Plato), vol. 2 (Hegel and Marx) (London).

—— (1959), *The Logic of Scientific Discovery*, (London).

—— (1976), *Unended Quest: An Intellectual Autobiography* (London).

Porter, T. M. (1986), *The Rise of Statistical Thinking: 1820–1900* (Princeton).

Prideaux, G. J. et al. (2010), 'Timing and Dynamics of Late Pleistocene Mammal Extinctions in Southwestern Australia', *Proc. Natl. Acad. Sci. USA*, 107: 22157–22162.

Prichard, J. C. (1813), *Researches into the Physical History of Man* (London).

Raven, C. E. (1986), *John Ray: Naturalist*, repr. (Cambridge).

—— (1947), *English Naturalists from Neckham to Ray: A Study of the Making of the Modern World* (Cambridge).

Ray, J. (1691), *The Wisdom of God Manifested in the Works of the Creation* (London).

Reanney, D. C. (1974), 'On the Origin of Prokaryotes', *J. Theor. Biol.*, 48: 243–251.

Reeves, H. et al. (1999), *Origins: Cosmos, Earth, and Mankind* (New York).

Robertson, M.P. and Joyce, G.F. (2012), The Origins of the RNA World, *Cold Spring Harb. Perspect. Biol.*, 4: a003608.

Romiguier, J. et al. (2013), 'Less Is More in Mammalian Phylogenomics: AT-Rich Genes Minimize Tree Conflicts and Unravel the Root of Placental Mammals', *Mol. Biol. Evol.*, 30: 2134–2144.

Roy, S. W. (2006), Intron-Rich Ancestors, *Trends Genet.*, 22: 468–471.

Rule, S. et al. (2012), The Aftermath of Megafaunal Extinction: Ecosystem Transformation in Pleistocene Australia, *Science*, 335: 1483–1486.

Sakamoto, M. et al. (2016), Dinosaurs in Decline Tens of Millions of Years before Their Final Extinction, *Proc. Natl Acad. Sci. USA*, 113: 5036–5040.

Sanchez, G. et al. (2014), Human (Clovis)–Gomphothere (*Cuvieronius* sp.) Association ~13,390 Calibrated yBP in Sonora, Mexico, *Proc. Natl Acad. Sci. USA*, 111: 10972–10977.

Sandler, I., and Sandler, L. (1985), A Conceptual Ambiguity That Contributed to the Neglect of Mendel's Paper, *Hist. Phil. Life Sci.*, 7: 3–70.

Schulte, P. et al. (2010), 'The Chicxulub Asteroid Impact and Mass Extinction at the Cretaceous-Paleogene Boundary', *Science*, 327: 1214–1218.

Sereno, P. C. et al. (2009), 'Tyrannosaurid Skeletal Design First Evolved at Small Body Size', *Science*, 326: 418–422.

Skoultchi, A. I., and Morowitz, H. J. (1964), 'Information Storage + Survival of Biological Systems at Temperatures Near Absolute Zero', *Yale J. Biol. Med.*, 37: 158–163.

Slack, K. E. et al. (2006), 'Early Penguin Fossils, Plus Mitochondrial Genomes, Calibrate Avian Evolution', *Mol. Biol. Evol.*, 23: 1144–1155.

Stigall, A. L. (2012), 'Speciation Collapse and Invasive Species Dynamics during the Late Devonian "Mass Extinction"', *GSA Today*, 22: 4–9.

Stringer, C. B., and Andrews, P. (1988), 'Genetic and Fossil Evidence for the Origin of Modern Humans', *Science*, 239: 1263–1265.

Strobel, S. A. (2013), 'Metal Ghosts in the Splicing Machine', *Nature*, 503: 201–202.

Sullivan, R. M. (1987), 'A Reassessment of Reptilian Diversity across the Cretaceous-Tertiary Boundary', *Contributions to Sci.*, 391 (Natural History Museum of Los Angeles County).

Tishkoff, S. A. et al. (2009), 'The Genetic Structure and History of Africans and African Americans', *Science*, 324: 1035–1044.

Thorne, A. G., and Wolpoff, M. H. (1992), 'The Multiregional Evolution of Humans', *Sci. Am.*, 266(4): 28–32.

Toffano-Nioche, C. et al. (2013), 'RNA at 92°C', *RNA Biology*, 10: 1211–1220.

Towns, D. R. et al. (2007), 'Responses of Tuatara (*Sphenodon punctatus*) to Removal of Introduced Pacific Rats from Islands', *Conserv. Biol.*, 21: 1021–1031.

Tyson, E. (1699), *Orang-Outang, sive Homo Sylvestris: Or, the Anatomy of a Pygmie Compared with That of a Monkey, an Ape, and a Man*, repr. in 1973 (London).

Von Appen, A. et al. (2015), 'In Situ Structural Analysis of the Human Nuclear Pore Complex', *Nature*, 526: 140–143.

Vigilant, L. et al. (1991), 'African Populations and the Evolution of Human Mitochondrial DNA', *Science*, 153: 1503–1507.

Wanless, R. M. et al. (2007), 'Can Predation by Invasive Mice Drive Seabird Extinctions?' *Biol. Lett.*, 3: 241–244.

Watson, D. M. S. (1929), presidential address to zoology section, British Association for the Advancement of Science, in R. A. Fisher and J. H. Bennett (eds.), *The Genetical Theory of Natural Selection: A Complete Variorum Edition* (Oxford).

Watson, E. E. et al. (2001), '*Homo* Genus: A Review of the Classification of Humans and the Great Apes', in P. V. Tobias et al. (eds.), *Humanity from African Naissance to Coming Millenia* (Florence), 307–318.

Watson, J. D, and Crick, F. H. C. (1953a), 'Molecular Structure of Nucleic Acids—A Structure for Deoxyribose Nucleic Acid', *Nature*, 171: 737–738.

Whewell, W. (1832), review of Charles Lyell's *Principles of Geology*, in the *London Review* (p. 126).

White, A. W. et al. (2010), 'Megafaunal Meiolaniid Horned Turtles Survived until Early Human Settlement in Vanuatu, Southwest Pacific', *Proc. Natl Acad. Sci. USA*, 107: 15512–15516.

Wilf, P. et al. (2004), 'Correlated Terrestrial and Marine Evidence for Global Climate Changes before Mass Extinction at the Cretaceous-Paleogene Boundary', *Proc. Natl. Acad. Sci. USA*, 100: 599–604.

Wilkins, A., and Holliday, R. (2009), 'The Evolution of Meiosis from Mitosis', *Genetics*, 181: 3–12.

Wilson, A. C., and Cann, R. L. (1992), 'The Recent African Genesis of Humans *Sci. Am.*, 266(4): 22–27.

Wilson, G. P. et al. (2012), 'Adaptive Radiation of Multituberculate Mammals before the Extinction of Dinosaurs', *Nature*, 483: 457–460.

Winchester, W. (2001), *The Map That Changed the World: The Tale of William Smith and the Birth of a Science* (London).

Wragg, G. M. (1995), 'The Fossil Birds of Henderson Island, Pitcairn Group—Natural Turnover and Human Impact: A Synopsis', *Biol. J. Linn. Soc.*, 56: 405–414.

Worthy, T. H., and Clark, G. (2009), 'Bird, Mammal and Reptile Remains', in G. Clark and A. Anderson (eds.), *The Early Prehistory of Fiji* (Canberra), 231–258.

Yusupov, M. M. et al. (2001), 'Crystal Structure of the Ribosome at 5.5: A Resolution', *Science*, 292: 883–896.

Zhong, B. et al. (2014), 'Two New Fern Chloroplasts and Decelerated Evolution Linked to the Long Generation Time in Tree Ferns', *Gen Biol. Evol.*, 6: 1167–1173.

Zubkov, M. V., and Tarran, G. A. (2008), 'High Bacterivory by the Smallest Phytoplankton in the North Atlantic Ocean', *Nature*, 455: 224–226.

Printed in the United States
By Bookmasters